Research for Development

The series Research for Development serves as a vehicle for the presentation and dissemination of complex research and multidisciplinary projects. The published work is dedicated to fostering a high degree of innovation and to the sophisticated demonstration of new techniques or methods.

The aim of the Research for Development series is to promote well-balanced sustainable growth. This might take the form of measurable social and economic outcomes, in addition to environmental benefits, or improved efficiency in the use of resources; it might also involve an original mix of intervention schemes.

Research for Development focuses on the following topics and disciplines:

Urban regeneration and infrastructure, Info-mobility, transport, and logistics, Environment and the land, Cultural heritage and landscape, Energy, Innovation in processes and technologies, Applications of chemistry, materials, and nanotechnologies, Material science and biotechnology solutions, Physics results and related applications and aerospace, Ongoing training and continuing education.

Fondazione Politecnico di Milano collaborates as a special co-partner in this series by suggesting themes and evaluating proposals for new volumes. Research for Development addresses researchers, advanced graduate students, and policy and decision-makers around the world in government, industry, and civil society.

THE SERIES IS INDEXED IN SCOPUS

More information about this series at http://www.springer.com/series/13084

Paolo Pileri · Rossella Moscarelli
Editors

Cycling & Walking for Regional Development

How Slowness Regenerates Marginal Areas

Springer

Editors
Paolo Pileri
Dipartimento di Architettura e Studi Urbani
Politecnico di Milano
Milan, Italy

Rossella Moscarelli
Dipartimento di Architettura e Studi Urbani
Politecnico di Milano
Milan, Italy

ISSN 2198-7300 ISSN 2198-7319 (electronic)
Research for Development
ISBN 978-3-030-44002-2 ISBN 978-3-030-44003-9 (eBook)
https://doi.org/10.1007/978-3-030-44003-9

This Springer imprint is published by the registered company Springer Nature Switzerland AG
The registered company address is: Gewerbestrasse 11, 6330 Cham, Switzerland

Introduction: Slowness as a Paradigm of Regional Development

Aims of the Volume

There is no great crisis without everyone seeking some kind of *development* to cling on to get through it. But what kind of *development*? This seems to us to be the great question to ask. Not all are neutral in relation to environmental impact. Not all generate opportunities where they are needed. Not all are capable of distancing themselves from the rhetoric that has led to crises of recognizing them to avoid them in the future. This exercise, as necessary as it is complicated, is made even harder in the so-called marginal areas, whose name keeps them at a distance compared to centers, the beating hearts where everything is available. In marginal areas, consequently, this is what people often say: “there is nothing”, “everything is at a standstill”, “it’s beautiful, but I’d never live there”, and “everything is terribly slow”. Of all characteristics, slowness seems to be the one most capable of condemning marginal areas to never seeing a future of redemption, and therefore it is the hardest to eradicate, but at the same time it is the most urgent of all. In this volume, we therefore want to try inverting the negativity of the combination of marginal areas—slowness. Slowness, if seen from a different perspective, is, on the contrary, a strategic resource from which it is possible to start in order to design a concrete future for those fragile territories, as has already happened along the Danube in Austria and Germany, along the Camino de Santiago in Spain, and in other European cases that will be discussed.

The intent of the volume is therefore to present the existence of a slowness that is the key to open all the doors to places that are in any case considered, regardless of them being central or peripheral. To unhinge the split between A- and B-list places, slowness is a tool that has much to teach about urban practice and planning, although to date it has had little space on the agenda. The study we propose here is relative to a particular practice of slowness: the going, the traveling, the moving along a line that crosses territories and not only inside the cities. The thought of a slow movement along a line is well represented by a walk or a bike ride. And if we think about a walk that has the length of a journey and not a day trip, then we are thinking about pathways and bike paths.

Within the volume we have brought these elements to the table: marginal areas, the planning and policies that support these areas, slowness, and travel (not simply tourism). We change things up and wonder what could be born from the hybridization of worlds that do not dialogue because of system rigidity, thus negating the ignition of possible sustainable developments that are very interesting and appropriate to marginal areas. By doing so, we discover alternatives to the beaten paths. We understand that focusing in depth on the lesson of slowness, in our accelerated model, can truly change the paradigms of territorial planning. A bicycle route or a pathway, with their linear development that crosses the territory overcoming the formalism of administrative boundaries and fragmented public management that only obey bureaucratic criteria, can truly challenge the relationship between places and planning, improving it. A planning that no longer has the administrative limitation or frame of reference of the municipality's, the Region's, or of the areas that are considered marginal and therefore requiring a "special" planning, but which bases its development expectation on a live trace, experienced by walkers and cyclists, who do not care about boundaries but about the beauty of what they are passing through. The meaning of planning and of an original strategy starting from the project of a bicycle route or a pathway comes, therefore, from regions defined by the line itself and not by the administrative structures the planning depends on. This inversion designs new geographies and opens new opportunities, even economic and social, freeing up energy and projects that so far have been unable to take off because the run-up was too short.

Structure of the Book

The volume is divided into four parts.

Part I lays the theoretical basis on which the main themes of the book are developed: territorial marginality with its development policies, slowness, and the journey. Chapter "From Slow Tourism to Slow Travel: An Idea for Marginal Regions", by the editors, introduces a consideration of how slowness, applied to tourism and journey, can be a tool for territorial development, in particular, for marginal areas. The meaning of tourism is explored critically. In chapter "The Seat and the Saddle: How Slow is Quick and Fast is Stuck", anthropologist Tim Ingold explains how a simple experience of slowness, such as pedaling on a bicycle, allows an observation of the world that is totally different from what happens from the seat of a fast car. Rossella Moscarelli, in chapter "Marginality: From Theory to Practices", presents a consideration on territorial marginality and how, depending on the different meanings of "margin", different strategies for interventions can be derived. Having introduced the topic of territorial marginality, the following chapters tackle the topic from the perspective of public policies. Manos Matsaganis, in chapter “Marginalised Areas as a Public Policy Concern”, tackles this on an international scale, discussing some key elements of the matter, such as territorial divergence, on the basis of marginalization processes. Benedetta Silva, in chapter “Italian Policies on Marginal Territories: An Overview”, presents on the other hand a brief history of Italian public policies that since the 50s to date have tried to reduce the territorial unbalance suggesting extraordinary interventions.

Part II focuses on the topics presented in the first part, interpreted within the specific Italian context. In chapter “Slowness to Discover the Ordinary Italian Landscape”, Chiara Visentin talks about slowness in relation with the Italian landscape, showing how moving slowly within allows us to discover the common and “ordinary” beauty of places, in opposition to the spasmodic search for the exceptional. Discovering the territory, at the right speed, can also be the tool that allows us to read the history of food, as Massimo Montanari in “Food on the Bike” explains. In chapter “The Role of Historic Roads to Preserve and Valorize the Landscape”, Alberta Cazzani and Maurizio Boriani demonstrate how pathways and historical roads, from which it is possible to plan slow territorial lines, are themselves a form of cultural heritage, a trace that can revisit the development itself of the territory. Finally, Alessandra Oppio and Marta Dell’Ovo, in chapter “Cultural Heritage Preservation and Territorial Attractiveness: A Spatial Multidimensional Evaluation Approach”, present an Italian case study that shows the potential for regeneration and recovery of certain public ownership buildings, starting from bike routes and pathways. This introduces the reader to a further dimension of recovery and planning of services along the slow lines.

Part III discusses in depth VENTO bicycle route, the 700-km-long bicycle tour ridgeline created by a research group of the Politecnico di Milano which has become a symbolic case study of a project of the first long bicycle tour line in Italy. In chapter “Slow Travel Project for Enhancing Territories: Motivations and Directions” Paolo Pileri, creator of VENTO and scientific Coordinator of the research group, explains the reasons for the project. To follow, in chapter “Design, Public Engagement and Communication: Reframing Methodology”, Alessandro Giacomel, Diana Giudici, and Camilla Munno of the VENTO research group demonstrate the research methodology that allowed them to put the idea into practice throughout the various work stages. Developing the project of a bicycle route that aimed to be a project of territorial regeneration of marginal areas has required a mixed research approach, made up of a collection of case studies, publications and communication at various levels, technical planning, local involvement, and institutional cooperation. Lastly, Catherine Dezio, in chapter “Narration of Cultural Heritage as *Antifragile* Tool”, discusses the narrative and educational value of a bike route such as VENTO, designed to create an intimate connection with the territory it crosses.

Part IV looks at interesting European cases where bicycle routes and pathways have allowed for an economic development, and not only, of the areas that were crossed. The goal is to bring out good practices especially regarding public policies, its various scales that have set off, promoted, and sustained processes of construction and development of the main lines of slow tourism in Europe. In chapter

"Learning from Experience: A Set of European Policies", Federica Bianchi proposes a critical review of five European countries: The Netherlands, Germany, France, Denmark, and Switzerland, who have worked on national cycling plans, highlighting the process behind such strategies. To follow, Xavier Brice, in chapter "The United Kingdom's National Cycle Network: Paths for Everyone, Past, Present and Future", explains the English case and of Sustrans, the company that has promoted the National Cycle Network. In chapter "The Singularity of the Camino de Santiago as a Contemporary Tourism Case", Ruben Lois and Lucrezia Lopez present the famous case of the Camino de Santiago, in Spain, tracing the historical reasons for its success and the urban and territorial success that was activated along the line of the Camino. Michael Meschik, in chapter "The Success of the Cycle Tourist Backbone Along the Danube in Germany and Austria", finishes by discussing the case of the bicycle route along the Danube, from Vienna to Passau, that, especially in the Austrian and German stretch, proves to be one of the most important lines of bicycle tourism in Europe.

Paolo Pileri
paolo.pileri@polimi.it

Rossella Moscarelli
rossella.moscarelli@polimi.it

Contents

A Proposal of Regeneration for Marginal Areas

From Slow Tourism to Slow Travel: An Idea for Marginal Regions

Paolo Pileri and Rossella Moscarelli

Abstract Tourism represents a strategy that can be used to imagine the development of marginal areas. But what type of tourism? Slow tourism is considered in literature and by the main development policies of marginal areas as one of the forms of tourism that best lends itself to the specific context of these territories. There are three factors whose possible relationship and interaction will be studied: tourism, slowness, and marginal areas. In this piece, the matter of marginal areas is not discussed, and it is taken as fact. What is discussed is the combination of slowness and tourism, often identified with the idea of "slow tourism". The article proposes its own definition of slow tourism, where slowness, as a conscious and alternative attitude, invests in and modifies the economic sector of tourism. We therefore identify the attitudes of slowness that bring meaning to a territorial project, useful to the development (not only financial but also cultural and social) of marginal areas. From tourism, we move on to travel, a free and discovery-based approach, in line with the lessons that slowness can provide.

1 Which Tourism to Regenerate Marginal Areas?

Tourism has been considered as a key strategic sector for development via the European Structural and Investment Funds (COM 2010). In an interesting volume elaborated by OECD (2006), tourism has been seen also as a tool to enhance the growth

---

[1]In this article, we will not consider what constitutes a marginal region (which will be discuss in other essays in the volume). We will therefore talk about "less developed", "marginal", "peripheral", "fragile", and "inner" (referring to the Italian context) areas to indicate the regions that for many different reasons have been included in recent support and development policies (rural, isolated, mountainous, shrinking areas, etc.).

P. Pileri · R. Moscarelli (✉)
Politecnico di Milano, Milan, Italy
e-mail: rossella.moscarelli@polimi.it

P. Pileri
e-mail: paolo.pileri@polimi.it

P. Pileri and R. Moscarelli (eds.), *Cycling & Walking for Regional Development*,
Research for Development, https://doi.org/10.1007/978-3-030-44003-9_1

of many European regions, in particular, the less developed and marginal regions,[1] due to its considerable spillover and potential for job creation, particularly for young people. Actually, especially for those areas where traditional jobs, such as agriculture, livestock, and handicraft were gradually abandoned, tourism can become the most important sector (Pelc 2018). However, we cannot stop wondering about which tourism can do so in a sustainable way and with which principles and goals.

The combination of territorial marginality and tourism has been widely studied both in the academic debate and within the public policies that are proposing development strategies for fragile areas. Tourism, as a matter of fact, is seen as a way to enhance and exploit the territorial capital and it is proposed as a valid strategy to revitalize marginal territories (Espon 2017). Also in the most recent Italian policy regarding this, the National Strategy for Inner Areas, tourism is one of the five lines of local development, together with "*active territorial/environmental sustainability protection, valorization of agriculture and food system, activation of renewable energy supply chains and know-how and crafts*" (UVAL 2014: 21). In the Italian debate, the most recent attempts to study tourism as a trigger for the development of fragile areas was carried out by Meini (2019) and Ferrari and Cavuta (2018). Actually, the idea that tourism can help in the development of marginal areas is not new, also applied to the context of the Italian inner areas (Formez 1980). At the same time, already in the past, there was some skepticism about the idea that tourism could be the panacea of all the problems of marginal areas. In this respect, Manlio Rossi Doria, the Italian father of the inner area concept, wrote:

> We must therefore muster the courage to look at the mountain's issues with new eyes, laws and institutions. Only in this way will they be solved. Everything else is an illusion, at the bottom of which is abandonment through the landslide of inexorable exodus. Illusion, in particular, is indeed the fig-leaf that is much discussed: mountain tourism. Tourism is, of course, a valuable thing, that in some cases will allow a certain development of privileged areas, but for most of the southern mountain area it will carry a very limited and defined weight. [...] We must recognize that, except in few cases, tourism is not a solution but an illusion.
>
> (Rossi Doria 1982/2003: 98)

In sum, Rossi Doria points out that tourism cannot be the unique solution for all the problems affecting marginal areas. To this, we must add that, even where tourism is an effective sector of success and economic growth, the potentially negative impact must not be overlooked. The concept of sustainability related to tourism, which gives life to so-called "sustainable tourism", can seem like an oxymoron, as tourism, an activity that exploits resources, and can be not considered completely excluded from the negative impact on the territory (environmental, economic, social, and cultural impact). In a recent publication, Pelc and Koderman (2018) have collected several essays in order to discuss the positive effects of tourism in marginalized context and even the risks to take into account. In truth, in their introduction, they present the ambiguity of the role of the tourism, by explaining that

> Tourism as such seems to be very promising and often imposes big expectations in areas with developmental problems. Especially in backward rural areas with relatively unspoiled nature and poor accessibility to industrial, commercial and residential centers, tourism is often seen as a redeeming developmental activity which may be highly overrated. Tourism cannot be a substitute for uncompetitive agriculture or missing industry. There are only few special places where the majority of the local population can live from the revenues from tourism. Not every single rural area can expect such an outcome from tourism development. However, tourism can play an important role in regional development when it is combined with the development of other sectors. (Pelc and Koderman 2018: 22)

This awareness suggests, therefore, that every regeneration proposal that relates to tourism will have to always establish a threshold that cannot be overcome, for territories to live on, although not completely, on tourism. In this respect, Marco D'Eramo affirms

> Until the influx of visitors exceeds this limit, tourists use services and benefits that are designed for residents. Beyond this threshold, on the other hand, residents are forced to use services that were developed for tourists. (D'Eramo 2017: 78)

Tourism is therefore a potential sector for the development of marginal areas (and more). But it must be handled with care. An interesting paper by Salvatore et al (2018) examines the relationship between what they call the rural peripherality and the tourism transition in the Italian context. They develop a critical analysis of the role that contemporary tourism can play in peripheral rural contexts as a driver of change. In this study, the authors see in new forms of locally based tourism, alternative to mass tourism, an opportunity for peripheral areas. The interesting thing here is that they speak not about tourism in general, but rather they specify the typology of tourism that can be suitable for the fragile territories. This seems to suggest that marginal areas, because of their nature and their fragile condition, need a specific consideration and a model of development that is consistent with their characteristics. In this sense, the question is which type of tourism, as a possible model of development, is more appropriate to the marginal area's context. In this article, a reflection on so-called slow tourism, which we will define, is proposed. The idea that slow tourism could be suitable to develop marginal areas has already been investigated both in Italian (Manella 2017; Nocifora et al. 2011) and in foreign literature (Matos 2004). All these studies have in common the idea that marginal territories, more than others, seem to be destined to slowness, offering a nonconformist type of tourism. These territories, described as "slow" (Lancerini 2005), can become icons of beauty and good quality of life. In this respect, the slowness of these places, from a negative meaning, can be turned into care for the environment and the landscape, a high quality of life, a little known historical and architectural heritage of quality, a strong local identity, a taste of local quality products.

2 Slow Tourism: A Collection of Definitions

The idea of so-called slow tourism is quite well cited. Nevertheless, taking for granted its meaning is a mistake and a review of the literature can easily demonstrate that there is no shared definition about what "slow tourism" is. First of all, slow tourism and slow travel are usually confused, even if there are some differences between them as we will see in the next paragraph. Moreover, depending on the aspect we decide to emphasize, we will find different ways to intend what slow tourism is. We have identified three different approaches by which slow tourism or slow travel have been defined: (1) in opposition to traditional and mass tourism; (2) as tourism with low-carbon impact; and (3) as an experience where the most relevant aspect is the authenticity of the journey.

Starting from the first, several studies focus on the idea that slow tourism is an attitude, an ethical choice, in opposition to traditional tourism, the so-called "mass tourism". The most influential exponent of this theory was Krippendorf (1987). In his text, The Holiday Makers, he expressed his doubts about the sustainability of mass tourism. Many of the approaches suggested by Krippendorf over 25 years ago (such as holidays at home or nearer to home, an emphasis on locality, and diversity) could be included in the recent idea of slow tourism and travel. Slow tourism, as a matter of fact, is usually seen in contrast with the negative externalities of mass tourism, which is characterized by the extensive structural and infrastructural development of a territory based principally on economic interests, with an adequate consideration of environmental and social factors. As McGrath and Sharpley (2017) explain, the slow travel narrative focuses on the so-called "art of travel" (De Botton 2003), in opposition to the loss of such "art" in the case of "mass tourism".

We have collected in Table 1 some definitions that we have found in line with this approach.

The second approach emphasizes that the mode of transport should have lower environmental impact and less travel, the main features of slow tourism. This aspect focuses on alternatives to air and car travel, such as trains, busses, cycling, walking, boat both to and within a destination, and where travel to and from a destination becomes part of the holiday. From this perspective, slow tourism is seen as an environmental choice in order to reduce the huge amount of carbon emissions related to the traditional tourism sector (Dickinson and Lumsdon 2010). From this perspective, slow tourism excludes the three main transport modes associated with contemporary tourism development: the car, the cruise liner, and the airplane. In addition, this approach is also related to the choice of destination that should be close to home or reachable by low-carbon methods of transport. This appears in contrast with tourism travel distance that recently has grown as travel speeds have increased and travel costs decreased.

In Table 2, some definitions of slow tourism/travel according to this second view.

Finally, the third approach defines slow tourism mainly as a tool to experience the territory, where visitors engage in a deeper experience of place. In this sense, slow tourism is mainly an experience of the territory that involves people, services, and

Table 1 Some definitions where slow tourism is seen as an alternative to mass tourism. In all these definitions, it is possible to note the use of the words "new", "alternative", and "different". They highlight how the slow tourism approach is in opposition to traditional tourism

Author/s	Definition
Calzati and De Salvo (2017:40)	Slow tourism, together with other new forms of tourism such as ecotourism, rural tourism, etc., "*should be directed to overcoming the difficulties in relation to the inadequate economic, social and environmental sustainability typical of mass tourism*"
Blanco (2011: 127)	"*In the tourist field, slow tourism emerges as an alternative to the generalised form of consumption holiday. This alternative covers all market players, both consumers as producers […]. In other words, Slow tourism requires the responsibilities of all market players and goes through a transformation of the dominant imaginary*"
Heitmann et al. (2011)	The most significant principle in slow tourism is attributable to a different concept of vacation, which is no longer characterized by the number and quantity of experiences, but it is distinguished by living fewer experiences of higher quality
UNWTO (2012: 24)	"*Slow tourism allows a different set of exchanges and interactions than those available in the hurried context of mainstream tourism, with economic benefits to the host and cultural benefits to the tourist*"

Table 2 Definitions where slow tourism is seen as low-carbon impact tourism

Author/s	Definition
Dickinson et al. (2010: 482)	"*Slow travel is an emerging conceptual framework which offers an alternative to air and car travel, where people travel to destinations more slowly overland, stay longer and travel less*"
Dubois and Ceron (2003)	Slow tourism (train transport) and soft mobility (biking, walking) can be tools to image a low-impact tourism utopia
Mintel (2009: 1)	Slow travel is "*a trip made using non-aviation methods for departure*"

locations (Westwood et al. 2006). Moving slowly in the territory, traveling slowly, implies immersion in the environment, where participants become physically a part of the experience itself. In this sense, slow tourism can be read under the lens of Pine and Gilmore's (1999) matrix of experience realms.

It is not surprising that one kind of tourism that seems to be correlated to the idea of slow tourism is the "experiential tourism". Edgell and Swanson (2019) report an interesting definition made by the organization "Nature and Outdoor Tourism Ontario" (Canada), namely:

> Experiential tourism has become the current term that encompasses a variety of tourism and traveler categories […] where activities are environmentally sensitive, displaying respect for

Table 3 Definitions where slow tourism is seen as experience of the territory

Author/s	Definition
Gardner (2009: 11)	*"Slow travel also reshapes our relationship with places, encouraging and allowing us to engage more intimately with communities through which we travel"*
Heitmann et al. (2011: 117)	Slow tourism is *"a form of tourism that respects local cultures, history and environment, and values social responsibility while celebrating diversity and connecting people (tourists with other tourists and host communities)"*
Dickinson et al. (2011: 285)	Slow travel allows us to live *"the importance of the travel experience to, and within, a destination, engagement with the mode(s) of transport, associations with slow food and beverages, exploration of localities in relation to patrimony and culture at a slower pace"*

> the culture of the host area and looking to experience and learn rather than merely stand back and gaze. Experiential tourism involves active participation, involvement, even immersion.

Table 3 shows some definitions of slow tourism/travel where the main aspect is the experience of the place.

3 Slow Tourism or Slow Travel?

As we touched on earlier, when considering the difficulty in defining slow tourism the difference between slow tourism and slow travel must also be pondered. Based on how the "slow" experience is created, during the journey or in the place of arrival, we can distinguish, respectively, slow travel and "slow tourism". Leed (1992) helps us to better understand the issue by introducing the three stages of the journey, well described by Lavarini and Scramaglia:

> The first stage is the departure. The stage of abandonment of what we know and cherish to face the unknown. The second is the passage, meaning the path that serves as a link between the act of departing and the act of arriving. The third is, precisely, the arrival, or, in other words, when we try to integrate with the place of destinations. [...] According to Leeds, the cornerstone of the travel experience is the passage, toward which less and less interest is shown nowadays, gradually reducing itself to a merely symbolic aspect of many journeys. (Lavarini and Scramaglia 2017: 79)

Even in what concerns the slow attitude of journeys there are some doubts regarding the most important phase of the journey, whether the slow experience happens during the passage or the arrival. According to Lavarini and Scramaglia (2017), traveling slowly can allow us to rediscover the passage phase of a journey, to live the territory that is being crossed and to engage with the local culture with numerous activities. Slowness turns every journey into a journey of discovery, as Bodei (2009) defines it, and this helps us reveal the nature and history of things that meet, establishing a strong relationship between the I and the landscape. This vision can be particularly interesting especially in the case of marginal areas that are often completely bypassed by fast destination tourism where, indeed, the final destination is the objective and not the journey itself. On the contrary, travel can be seen as integral to the tourist experience, and, in some instances, it might be the main purpose (Lumsdon 2006). Moreover, the idea of tourism implies a consumption where people, the tourists are consumers. On the contrary, travel can be seen more as an experience, where people are not seen only as consumers, rather as travelers or, as Tim Ingold (2007) states, as wayfarers that are discovering the territory. As McGrath and Sharpley (2017) note, within the slow tourism[2] concept, the focus shifts away from the mode of transportation used and distance traveled to the product being supplied and consumed.

Even if slow travel and slow tourism are usually used as synonyms (Dickinson and Lumsdon 2010; Fullagar et al. 2012), we decide to assume the idea of the slow travel rather than tourism, it being more in line with the idea of experience rather than consumption, of the journey rather than the final destination, of the travelers rather than the tourists.

4 The Slowness (H)as Value

At this point, is it appropriate to wonder why slow tourism, which we will, henceforth, call "slow travel", is at the center, more than other topics, of the consideration of this essay. Our interest starts from having seen slowness as a strategic resource to design regeneration for non-metropolitan territories, especially in the more peripheral areas, thanks to the power it has in allowing the discovery of places. And if slowness, contrary to speed whose only goal is to suppress distance, generates closeness (Bodei 2009), this discovery, this slow attitude toward travel can become a chance for development for those that are crossed and those it crosses. Slowness is therefore the form of energy that we are interested into design travel, not merely tourism, and it is appropriate for marginalized areas and not vice versa. If we invert the order of the factors, in addition the result will change in the face of every theory: from travel/tourism + slowness we have slowness + travel/tourism. Knowing how to look to slowness first and then to any form of travel or tourism can become the antidote, potential though never definitive, with which we can avoid or limit the consumerist tendencies typical of quick, hit-and-run practices.

[2]It is curious to note that the first definition of slow travel, made by Pauline Kenny in 2000, on the slowtravel.com website, does not incorporate the idea of the journey to a destination into slow travel. Indeed, the essence of slow travel, according to Kenny, is about staying in one place and then using the time to explore nearby localities, thus getting to know more about local culture and way of life. From this perspective, it is difficult to identify a distinction between slow tourism and other alternative (to mass) approaches to tourism development, such as ecotourism and, more generally, sustainable tourism.

This consideration is generated, therefore, by what slowness can teach tourism, especially in an age so taken over by the myth of speed. The proposal is to rehabilitate the concept of slowness from a synonym of backwardness (Rosa 2015) and waste (of time, money, energy) to the pivot of a possible, desirable, lasting, alternative model of development compared to a present where "everyday life is a sea that engulfs us with requests" (Rosa, 2015). Even in free time, this can happen if speed imposes a way to function in the world, which is made up of frenetic experiences, rapid and fixed successions, the need to fill every second with new and different facts.

To introduce any consideration on slowness, we must therefore face the fact that there is a speed limit under which we can define slowness, and over which there is no slowness. Two interpretations from different times and contexts aid us in this. The first, oldest, and international interpretation is by Ivan Illich and was written in his famous book Energy and Equity (1973), at the time of the first great energy crisis in modern times, a time when Illich understood that speed could be a discriminating factor that could limit social inclusion, contrary to slowness that did not present this limit.

> "Every time a public vehicle has gone beyond 25 km an hour equity has diminished, while the scarcity of time and space increased" (Illich 1973/2006: 19).

The second, more recent and Italian, is by Ermanno Rea, who is following a slow journey along the Po river, writes:

> Twenty kilometers per hour are an ideal speed, as things that fall within our gaze can appear both concrete and fake, material and immaterial. Twenty kilometers per hour are a sort of "golden" length of time for any touristic experience that desires both plastic certainty and archetypes, real things but also everything that seems to be hiding, and perhaps does indeed hide, behind real things. (Rea 1996/2017: 155).

From these two considerations, we can draw a limiting value of speed below which the area of slowness lies. 20–25 km/h is a speed that can easily be achieved and maintained physically or using a modest quantity of energy: on foot, by bicycle, canoe, small motor or sailboats, by horseback or pack-mule, and so on. If this is the speed limit and these are the possible forms of movement, we can now focus on what the advantages of slowness are.

1. Slowness has its trails

Low speed is still a speed, connected to a movement that makes its way through locations that belong to geographies. Every slow movement follows a trace or a line or a beam of lines that are themselves within a plan which, if we look closely, is the territory. As we will see, the lines can be short or long, wide or narrow, continuous or disconnected, sloping or flat, well or badly accessorized. The features of the trail influence the movement experience of our slow traveler. The lines can be ancient trails like medieval paths in Europe or natural trails like the banks of a river or specifically planned and created routes: paths, bike routes, tracks, channels, bridleways, or other types borrowed from what the region has to offer, such as dirt roads, embankments, sheep tracks, shores, cliffs, docks, headlands, and more. The ability to read, learn, and compose is one with slowness; we compose as we make our way on the path.

2. Slowness is a lens

In Italian, the world slowness, "*lentezza*", includes within it the word "*lente*", lens: the tool that magnifies everything we see, that makes what would otherwise elude us visible, and that allows us to focus on what would otherwise be out of focus. It is curious that this particular lens is contained within a slowed down word, almost as to say that its contrary, speed, does not allow us to capture the blurry backdrops and "patches of color", as Aldous Huxley (1989) called them. Slowness allows us to fully savor the complexity of the vision that is made up of feeling, selection, and perception. When moving in a hurry, the feeling we often call fleeting prevails, as it escapes without leaving anything. Slowness, instead, allows us to dwell on particulars that we can therefore learn and recognize (selection) to then interpret, through the cultural experience we gain, to the point of perceiving them completely. Slowness is therefore a teaching tool and mental energy because it feeds memory, a vitamin in turn for perception.

3. Slowness weighs nothing

Lightness is another specific feature of slowness. It relates to the low environmental impact of slowness related to the use practices, the service structures, and the infrastructures. Lightness is the reference principle to which slowness looks to for its environmental responsibility. If a path, a track, a host structure, a footpath, the waste management, or the food and water supply want to be part of a slowness project, they will also need to be light, in other words thought out and created in order to not weigh on the environment.

4. Inclusion and accessibility

Because of slow speed, the spectrum of people that can approach and practice this way of moving is broadened, as it does not require an expensive vehicle, nor particular abilities, high financial availability, competitive training, and costs to access the infrastructures. Slowness naturally incorporates an excellent potential for social inclusion, a quality that is highly interesting for a plan of territorial regeneration.

5. Travel "in between"

Slow, continuous, and linear movement feeds upon stories that are collected step by step, and which end up writing the plot of stories that are created on the way back. It's a continuous experience of vicinity with what is encountered. In the perspective of speed, the journey to reach a destination is seen as an obstacle and a waste of time. In many cases, transport is simply seen as an accessibility tool linking destinations to source markets (Prideaux 2000). For those who practice speed, slowness is a source of anxiety and a problem, as it is something that gets between them and the destination, both in the way there and back. They do not experience the journey; they endure it (Ingold 2007). It is clearly perceived as a travel cost, not a travel benefit. Goodwin and Frances (2003) find that slow travel could be interpreted as a form of ethical purchasing behavior. This involves applying additional criteria to the typical consumer behavior perspective. Indeed, travel is usually modeled as a cost to be minimized, both in terms of time and financial outlay. Based on such models, slow travel would not be seen as a rational decision, as it can take longer and cost more, especially in the current era of low-cost airlines.

6. The redemption of the time

The concept of slow is seen also as a reaction to the cult of speed. McGrath and Sharpley (2017) affirm that it is with the inherent fast nature of many Western cultures that leisure, unhurriedness, and slowness are becoming increasingly attractive concepts. The modernity of this present time has accustomed us to pressing on the accelerator, trying to fill every moment with more actions within a time unit, but without ridding us of that sense of time loss that Hartmud Rosa calls "time famine" that degrades and depresses us (Rosa 2015). Technology simply helps us deceive ourselves in thinking we are optimizing time that, despite this, is not as free as before. In slowness, this happens less or not at all, and time belongs to us again. Deciding what to do with our time goes back to being a great and satisfying form of power (Sepulveda 2013). During slow travel those who slow down win and those who speed up loose, contrary to what happens in urban geographies where speed is "the prevailing social norm" (Rosa 2015) and those who do not adapt to it are marginalized. This reconquering of time benefits every person who practices slowness, be it a traveler or a dweller, even in diverse ways, and it is therefore a precious material for a territorial project and a slow travel project we want to design.

5 Slow Travel: From Definition to Design

In our attempt to define slow travel, we realized that what we were thinking was more along the lines of a journey. A journey where the tools of slowness impose a rhythm that can generate benefits in those who practice it and also the places it is practiced in. All of this, however, brings us back to a very defined choice: designing slowness as a multi-goal practice that is capable of regenerating the more marginalized areas, the people who in time have detached from an intense experience of landscapes due to the habit of speed, and also a concept of mobility that is ever more bent to the idea of transport, more focused on destinations than on journeys. Practicing slowness allows us, in fact, to activate a secret and complex tie with memory, ours and of the places, to which a combination of speed oblivion is countered (Kundera 1995). Going slow does not mean lagging behind, as modernity's catchphrases try to convince us incessantly. On one hand, it means without a doubt resisting a world where everything is too fast (Sepulveda 2013), but also once more finding the intensity of memory, the fascination of reflection, the joy of observation, and the relationship with others and things. We do not intend slowness as a proposal for a decrease or reduction of our own rhythm at any cost, but as "an invitation to focus on the essential, to review our priorities, to listen to what they can tell us, on what matters to us, truly" (Pons 2019). Speed, on the contrary, is inversely proportional to the intensity of oblivion: we miss things, they blend together, they disappear, they do not stick to us, and they are impossible to memorize because our brain is a slow machine (Maffei 2014) despite its thrilling rapidity. We have no time to see what is around us, but only to move straight on. Speed consumes us and makes us consumers of the sole destinations that, within a logic of mass tourism, must multiply and impress us to increasingly satisfy us. The tourism that is generated is an alienating one that fails to create a relationship between us and the landscape rather than with scripts and which is very similar to a single philosophy of the ephemeral and of self-satisfaction. Slowness is on the other hand a rekindling even beyond itself (Christin 2019) as long as that beyond is legible and therefore understandable.

Designing slowness means adopting a philosophy that is the basis of an idea of travel that then ambitiously constitutes the foundation on which a proposal for development of the territory and the individual is perched. But from philosophy, one must progress to practice, if the aim that is moved by this consideration is to imagine a blueprint for development that is valuable to marginal areas. Designing slowness means creating a form to give it substance. This form is, in our view, represented essentially by a line: slow, long, continuous, in the open air and light. A line is to rehabilitate a travel philosophy where time is taken to discover unexpected beauty. A line materially keeps together, sews together, and unveils the many lost and dispersed beauties, giving them a voice, history, and stories, because *"giving shape to a length of time is the requirement of beauty, but it is also that of memory*" (Kundera 1995: 45). Line, beauty, memory added to which the fact that slowness needs to be pleasantly practicable, though with a certain amount of effort—effort helps us to focus—because the pleasure of a bicycle ride or a walk facilitates the reactivation of will that is the key to understanding the value of things that flow beside us, within the landscapes and the landscapes themselves. Understanding things requires overcoming a struggle of will, more than of the intellect. An understanding needs to be rid of the dust of banality and oblivion to go beyond the obvious that removes flavor from everything. But, again, to let go of the obvious it is necessary to undertake journeys of discovery (Bodei 2009). This is why, within the idea of the project, we are proposing in this paper, this set of four—line, beauty, memory, and pleasure—is grouped together like legs that support a chair, which in turn represents the territorial project within which the discovery journey exists; a journey which needs to happen freely (the back of the chair) taking advantage of a simple, clean, and accessible trace, where travelers stop and start wherever, and however their sensitivity takes them, not following a script. Due to all this, we find that we cannot define tourism or slow travel, but we can take a renewed interest toward a slowness that finds meaning and a regenerative role in the territorial designing process. This has convinced us to focus our attention first on the paradigm of slowness, as a potential activator of a proposal of territorial discovery, and only later of cycling or tourism, two potential implementations of this proposal. This proposal, in the form of a cycle path, a walkway, and any other type of path, cannot be carried out by chance; it needs a designing intent, a new territorial scale, technical knowledge and public objectives, policies and the definition of roles among contractors, citizens and associations, urban planning, and territorial government. Nor can it happen by using the old-fashioned rhetoric of transport or road engineering, as today, in order to plan a line, slow, open, and viable, various contributions of thought are required: humanistic, environmental, anthropological, esthetical, philosophical, artistic, etc. (Colville-Andersen 2018). The desire for movement, when it is slow, draws from other geographies and other sets of values that planners must take charge of and public stakeholders encourage. Slowness is therefore a great opportunity for architectural and territorial projects. Besides, as Francesco Careri explained, the act of walking itself has been of crucial importance in giving mankind the perception both of linearity and territorial structures (Careri 2006), as for years we organized our living space instinctively following shapes and tactics derived from the use of slowness.

Author Contributions The authors have shared the concept of the paper. P.P. has written the paragraphs 4 and 5. R.M. has written the paragraphs 1, 2, and 3.

References

Blanco A (2011) Una aproximacion al turismo slow. El turismo slow en las Cittaslow de España. Investigaciones Turisticas 1:122–133

Bodei R (2009) La vita delle cose. Laterza, Bari

Calzati V, De Salvo P (2017) Slow tourism: a theoretical framework. In: Clancy M (ed) Slow Tourism, Food and Cities. Pace and the Search for the "Good Life". Routledge, London, p 33–48

Careri F (2006) Walkscapes. Camminare come pratica estetica, Einaudi, Torino

Christin R (2019) Turismo di massa e usura del mondo. Eleuthera, Milano

Colville-Andersen M (2018) Copenhagenize. Island Press, Washington, The definitive guide to global bicycle urbanism

COM (2010) Commission communication 2010: Europe, the world's No 1 tourist destination - a new political framework for tourism in Europe. https://ec.europa.eu/growth/content/commission-communication-2010-europe-worlds-no-1-tourist-destination-new- political-0_en. Accessed 15 Sep 2019

De Botton A (2003) The art of travel. Penguin, London

D'Eramo M (2017) Il selfie del mondo. Feltrinelli, Milano

Dickinson J, Lumsdon L (2010) Slow travel and tourism. Earthscan, London

Dickinson J, Robbins D, Lumsdon L (2010) Holiday travel: discourses and climate change. J Transp Geogr 18(3):482–489

Dickinson J, Lumsdon L, Robbins D (2011) Slow travel: issues for tourism and climate change. Journal of Sustainable Tourism 19(3):281–300

Dubois G, Ceron J (2003) The interactions between climate change and tourism. In Climate Change, the Environment and Tourism: The Interactions, European Science Foundation – LESC Exploratory Workshop, Fondazione Eni Enrico Mattei, 4–6 June 2003, Milan

Edgell DL, Swanson JR (2019) Tourism policy and planning. Yesterday, today and tomorrow. Routledge, Abingdon

Espon (2017) Policy Brief: Shrinking rural regions in Europe. Towards smart and innovative approaches to regional development challenges in depopulating rural regions. https://www.espon.eu/rural-shrinking. Accessed 15 Sep 2019

Ferrari F, Cavuta G (eds) (2018) Turismo e aree interne. Esperienze, strategie, visioni. Aracne Editrice, Ariccia
Formez (1980) Progetto aree interne, vol. 1–7. Napoli
Fullagar S, Markwell K, Wilson E (eds) (2012) Slow tourism: Experiences and Mobilities. Channel View Publication, Bristol
Gardner N (2009) A manifesto for slow travel. Hidden Europe Magazine 25:10–14
Goodwin H, Francis J (2003) Ethical and responsible tourism: Consumer trends in the UK. Journal of Vacation Marketing 9(3):271–284
Heitmann S, Robinson P, Povey G (2011) Slow Food, slow cities and slow tourism. In: Robinson P, Robinson S, Dieke PUC (eds) Research themes for tourism. CAB International, Oxford, pp 114–127
Huxley A (1989) L'arte di vedere. Adelphi, Milano
Illich I (1973/2006) Elogio della bicicletta. Bollati Boringhieri, Torino
Ingold T (2007) Lines. A brief history. Routledge, London
Krippendorf J (1987) The holiday makers. Heinemann, London
Kundera M (1995) La lentezza. Adelphi, Milano
Lancerini E (2005) Territori lenti. Territorio, 34. Franco Angeli, Milano
Lavarini R, Scramaglia R (2017) Creative tourism as slow tourism. In: Clancy M (ed) Slow Tourism, Food and Cities. Pace and the Search for the "Good Life". Routledge, London, p 79–94
Leed EJ (1992) La mente del viaggiatore. Dall'Odissea al turismo globale, Il Mulino, Bologna
Lumsdon L (2006) Factors affecting the design of tourism bus services. Annals of Tourism Research 33(3):748–766
Maffei L (2014) Elogio della lentezza. Il Mulino, Bologna
Manella G (2017) Cittàslow. The Emilia Romagna case. In: Clancy M (ed) Slow Tourism, Food and Cities. Pace and the Search for the "Good Life". Routledge, London
Matos W (2004) Can slow travel bring new life to the Alpin regions. In: Weiermair K, Mathies C (eds) The tourism and leisure industry. Haworth, New York, pp 93–103
McGrath P, Sharpley R (2017) Slow travel and tourism: new concept or new label? In: Clancy M (ed) Slow Tourism, Food and Cities. Pace and the Search for the "Good Life". Routledge, London, p 49–62
Meini M (ed) (2019) Terre invisibili. Esplorazioni sul potenziale turistico delle aree interne, Rubbettino, Soveria Mannelli
Mintel (2009) Slow travel special report. London
Nocifora E, De Salvo P, Calzati V (eds) (2011) Territori lenti e turismo di qualità, prospettive innovative per lo sviluppo di un turismo sostenibile. Franco Angeli, Milano
OECD (2006) Innovation and Growth in Tourism. Organization for Economic Cooperation and Development, Paris
Pelc S (2018) Drivers of Marginalization from Different Perspectives. In: Pelc S, Koderman M (eds) Nature, Tourism and Ethnicity as Drivers of (De)Marginalization. Springer International Publishing, Insights to Marginality from Perspective of Sustainability and Development, pp 3–27
Pelc S, Koderman M (eds) (2018) Nature, Tourism and Ethnicity as Drivers of (De)Marginalization. Springer International Publishing, Insights to Marginality from Perspective of Sustainability and Development
Pine J, Gilmore J (1999) The experience economy. Harvard Business School Press, Boston
Pons N (2019) Perchè dire di no. In: Ferran F, Mattogno C, Metta A (eds) Coltiviamo il nostro giardino. DeriveApprodi, Roma
Prideaux B (2000) The role of the transport system in destination development. Tour Manag 21:53–63
Rea E (1996/2017) Il Po si racconta. Feltrinelli, Milano
Rosa H (2015) Accelerazione e alienazione. Einaudi, Torino
Rossi Doria M (1982/2003) Scritti sul Mezzogiorno. L'Ancora del Mediterraneo, Napoli

Salvatore R, Chiodo E, Fantini A (2018) Tourism transition in peripheral rural areas: Theories, issues and strategies. Annals of tourism Research 68:41–51
Sepúlveda L (2013) Storia di una lumaca che scoprì l'importanza della lentezza. Guanda, Milano
UNWTO (2012) Asia-Pacific Newsletter, Issue 27
Uval (2014) A strategy for Inner Areas in Italy: definition, objectives, tools and governance, Issue 31

The Seat and the Saddle: How Slow Is Quick and Fast Is Stuck

Tim Ingold

Abstract This brief essay reflects on what it means to go slow and fast. Drawing an analogy with a river, the waters of which run sluggishly near the banks but pick up speed in midstream, it contrasts both speed and slowness with the measure of velocity, calculated as the ratio of metric distance to chronological time, premised on the assumption that movement transports the traveller from one point to another, as from bank to bank across the river. This difference between going along (joining with the river current) and going across (taking the bridge) is linked to alternative modalities of perception, which depend on whether the traveller can maintain an upright posture with all-around vision, or whether their vision, from a reclining position, is oriented only forward. In the history of transport, this distinction is linked to that between the saddle and the seat. Re-entering the current of life means exchanging the seat for the saddle.

Imagine a river in its middle reaches, halfway between its source and the sea. Having already swelled from the confluence of multiple tributaries, and not yet relaxed into the lazy drift of its estuarine flow, it carries a mighty force. In the midst of the current, the waters bound along, swerving around obstructions, cascading over bluffs, thundering through ravines and heedless of everything in their path. Closer to the banks, however, in shallower waters, friction slows the current down, even to the point that its direction, in relation to the mainstream, is reversed. Trickles of water run here and there, feeling out every nook and cranny of the bank, investigating the weed, sometimes stagnating in little pools. Along the river, however, close to its banks on either side, are towns and villages inhabited by people. When they were first settled, the opportunities afforded by the river for travel by boat were a major factor in fixing their locations. At that time, long ago, travel overland was slow and hazardous. These days, however, boating on the river is largely done for pleasure. People travel by bicycle, car or public transport, along paved roads constructed for the purpose. In their working lives, they have turned their backs on the river, which

T. Ingold (✉)
University of Aberdeen, Aberdeen, UK
e-mail: tim.ingold@abdn.ac.uk

P. Pileri and R. Moscarelli (eds.), *Cycling & Walking for Regional Development*, Research for Development, https://doi.org/10.1007/978-3-030-44003-9_2

has now become more a barrier than a thoroughfare. Yet people have to get across, from one side to the other, in order to conduct their business, perhaps to commute from home on one side to offices or shops on the other. So bridges have been built, offering rapid transit from side to side. These movements, however, taking a direction orthogonal to that of the river's flow, are of a very different kind.

On the one hand there's the kind of movement that goes *along*. It is movement for its own sake, intrinsically forceful, with no fixed starting point or end point, but continually inflected by conditions in the landscape through which it passes. Such is the river current. On the other hand there's the movement that goes *across*. It has both a start point and an end point, and is driven by the purpose to get from one to the other. Whatever might transpire along the way is entirely incidental to this purpose, and counts as a distraction. Ideally, the traveller should focus single-mindedly on the intended destination, and not allow this focus to be lost to the lure of the intervening landscape. These two kinds of movement, moreover, offer radically alternative perspectives on life. Should we think of life as a continuous flow, the primary impulse of which is simply to carry on, making the most of what an environment affords and doing what it can to find a way through the obstacles thrown in its path? Or should we think of it, rather, as a career with milestones, a step-by-step progression from one achievement to the next, following the objectives of what professional counsellors and management consultants like to call a 'road-map'. Most of us, in practice, probably oscillate between the two perspectives. Though formally contradictory, they are not mutually exclusive. However, depending on which perspective you take, the other can show up differently. If you are in love with life, then career progression appears as a competitive rat race from which you yearn to escape. But in terms of the road map, the life that goes on in between is seen as recreation that helps to restore your competitive prowess.

To live we must of course move. But just as there are different perspectives on life and movement, so we can think differently of the means we use to get around. These include walking, cycling, and nowadays driving by car. Thus a walk or a bike-ride across the river bridge might be understood simply as a way to get from one side to the other, such as from home to business. But there's also a track along the river banks, which people like to take on a fine day. From your perspective on the bridge, these people—who appear to be going nowhere in particular—are engaged in a recreational activity, filling up intervals of 'free time' in a life otherwise regimented by the road map. Talk to the people themselves, however, and they might tell you that walking or cycling along the river bank makes them feel especially alive. The time it takes, for them, is time not *off* from life, but *of* its very essence. In their view it is the folk on the bridge, scurrying this way and that in their haste to cross, who are out of touch. Now in my home city, of Aberdeen in northeast Scotland, we have not only two rivers, with banks offering pleasant walks, but also a spectacular beach, where the city faces the North Sea. Citizens of all ages can be seen promenading along the esplanade that borders the beach. It's a Sunday morning, and I'm here to watch. Some people are alone, some move in pairs, or in family groups. No one, however, is en route from one place to another. The point is just to get fresh air and exercise, maybe to enjoy family time together.

Observing these promenaders, what immediately strikes me is the diversity of their means of locomotion. The promenade is closed to cars, but open to both walkers and cyclists. While this apparently offers a binary choice, between two feet and two wheels, my observations reveal a host of variations, in practice, in the combination of foot and wheel. In one family group, the parents are walking but their two small children, got up in gear and helmets, are on bicycles—the smaller child's fitted with stabilisers, as he's still not quite sure of his balance. The children keep going ahead, then they stop for their parents to catch up before shooting off again. It is great fun. In another group, an older person is being pushed along in a wheelchair. Then there's a young lad on a kick scooter: he has his left foot on the base-plate of the scooter, and uses the right to kick along. He doesn't really go much faster than the grown-ups he is with, but that's not the point. He simply enjoys the sensation, between each kick, of rolling along under his own momentum. The guy who clatters by on a skateboard, however, has picked up speed. For he is also clutching a sail. The wind, caught in the sail, powers him along without his even having to kick. He could almost be flying. So too with the pair of teenage girls who glide by on rollerblades. For them the sense of gliding along, of escaping the drudge of step-after-step, is liberating. It makes them feel tall. Then a young man comes hurtling along on a mountain bike, causing everyone else to scatter. I suppose he is just showing off.

For none of these people is the combination of feet and wheels taking them from A to B. They are not really going anywhere. And were you to track all the movements, the result would be a tangled and knotted braid, though the paving of the esplanade helps to keep it within certain margins. Many people also have their dogs with them, and if you were to factor in their tracks as well, the picture would become even more complicated. For besides the various combinations of feet and wheels, there's also the combination of two feet and four. Whereas on the beach the dogs run free, on the esplanade every animal is on a leash. It could be held in the owner's hand, or attached to a cycle handlebar. One cyclist goes by at a moderate pace, her dog running obediently alongside. As I watched, I was struck by how readily wheels of different sizes, from the smallest (as in rollerblades and stabilisers) to the largest (as in adult-sized bikes and wheelchairs), as well as feet in twos and fours, were incorporated into the moment-by-moment improvisations of a weekend promenade. Of course one could adduce many other examples from everyday life of foot-wheel combinations, from supermarket trollies and pushchairs to skateboards and suitcases on wheels, not to mention the rickshaw, a Japanese invention that remains ubiquitous in cities of East and South Asia. But it leads me to wonder why western city planners should have come to see wheeled and pedestrian movement as somehow opposed and incompatible, even to the extent of confining them to separate lanes or zones in the city. Could it be because of the priority overwhelmingly accorded to that four-wheeled engine of affluence, the car?

Cars, after all, have no use for feet, save to operate the controls. It is inconceivable that car wheels could join with feet, or run alongside them in the same procession, except on ceremonial occasions, in emergencies of mechanical failure, or—when stuck in mud or snow—all hands are needed to push! Thus, the more we take the car to be the prototypical instance of wheeled locomotion, the more we are inclined to think of feet and wheels as mutually exclusive. It is true that human bodies—like the bodies of most animals—are not designed for rotary motion. Our movements are of a reciprocal, back-and-forth, kind. That's why the invention of the crank, which converts reciprocal into rotary motion, was such a milestone in the history of the machine. The crank, of course, is common to both bikes and cars. On a bike it's the pedal, linking foot and wheel. On a car, by contrast, it is located within the engine, turning the thrust of the pistons inside the cylinders into revolutions for transmission to the axles. Not the crank, then, but the scale of automotive power puts the car in a league of its own. While a trained athlete can sustain about a third of the power of a horse, at most for an hour or two, an average car can produce the power of a hundred horses, or of three hundred people combined. It is no wonder that with such immense power behind the wheel, the car claims the highway to the exclusion of all else. For as countless accidents have proved to our cost, humans—even on two wheels—have as little chance in collision as an ant under the boot of a giant.

I was late in learning to drive. Until my mid-40 s I went everywhere by foot or bike, if not by public transport. I vividly remember the shock of travelling at what seemed to me high speed—almost 50 kilometres an hour! But the most difficult part of learning lay in suppressing my desire to look from side to side, to take in all the sights, sounds and curiosities along the way. Whenever something caught my attention, I would instinctively turn my head, only to be reprimanded by my instructor. To walkers and cyclists, unless racing with their heads down, this comes more or less automatically. When driving a car, however, the road turns into a visual tunnel. Your eyes stare straight ahead, using mirrors for side and rear vision. You are only too aware that a moment's distraction can be fatal. Again, it is the speed and power of the vehicle, rather than the wheels in themselves, that make driving so hazardous. Finding one's way at snail's pace through a traffic jam, or getting in or out of a confined parking spot, is an entirely different matter. Slow manoeuvres of this kind often call for a head that can swivel more than human anatomy allows. To look around, a turn of the neck is not enough; you have to turn your shoulders and trunk as well. For a walker, or for a cyclist, this is not a problem. Sitting upright in the saddle, the rider retains full freedom of upper body movement. The seat, however, is designed for a body that in the very act of sitting, surrenders to immobility. Reclining into the back-rest, strapped in across the chest and facing forward, it finds the execution of the turnaround both awkward and discomfiting.

Indeed the difference in movement and perception, epitomised in the distinction between seat and saddle, is profound—greater even than that between sitting down and walking erect. After all the saddle, originally an interface between mount and rider, makes it possible for both to move as one. But in the history of design, the antecedents of the seat lay not in the saddlery of the horse but in the furnishings of the carriage, wherein the passengers would sit, facing front or back, but oblivious to what was passing by outside. Whereas the saddle undergirds the rider, the seat holds its sitter in suspension, and the measure of comfort lies in the degree to which it insulates the body from the machinations going on below. The import of this difference, between seating and saddlery, may be greater even than that between bodily and automotive power, whether applied to two wheels or four. Many engine-powered machines, for example, are equipped with saddles rather than seats, including vespas, quad-bikes and agricultural tractors. In every case, the rider can enjoy uninterrupted, all-round vision. As a general rule, whereas the seat is part of the furnishings of an enclosed space, the saddle has the rider in the open, exposed to the elements—with at most a screen or awning to offer partial protection from wind, rain or intense sun—and in touch with the ground. Thanks to this direct sensory coupling, the rider is alert, in a way the driver can never be, to the sights, sounds and impressions of an environment which unfolds as he proceeds. And this heightened power of perception, I suggest, rather than any reduction in the ratio of distance travelled to elapsed time, is the hallmark of the kind of movement we are now inclined to call 'slow'.

The paradox of going slow is that movement and perception are actually quickened. They become more lively, nimble and alert to possibility. Here, distraction is a virtue, not a vice, for it allows you to attend to things that might otherwise flash by unheeded. What matters is not time but timing, not how long it takes to get from A to B but the ability to seize the moment—or what ancient Greek craftsmen knew as *kairos*—when, after perhaps a long wait, all the conditions needed for the accomplishment of a task are in favourable alignment. You have to be ready. Blink, and the moment's gone. Going slow means acting quick! What then becomes of its opposite? What does it mean to go fast? If slow is quick, then just as paradoxically, fast is stuck. It won't budge. To be made fast is literally to be fixed, tightly and immovably, into position. Why, then, should the speedy traveller go fast? He does so because to travel at speed you have to hold on tight, to surrender to the forces that bear you onwards, while bending them, through the slightest adjustments of the steer, in the direction you desire. Consider the rider of a powerful motorbike, more hunched over the saddle than sitting on it, head enclosed in a protective helmet, visor down. He's like a jet fighter pilot, in harness to a projectile. Only when he slows or comes to a stop can he afford to relax his grip. But then, does going fast—any more than going slow—depend on how long it takes? Do we go fast simply to get things done in as short a time as possible? If so, then to what purpose do we seek to lengthen our lives?

Fastness and slowness, on this account, are not degrees of velocity, nor can they be measured with a speedometer. They are rather experiential qualities of moving and perceiving that seek to break the stranglehold of chronological time and metric space imposed by the logic of the market. It is this logic that hurries and harries us, forcing us to race against the clock. Let us then return, in conclusion, to the image with which I began, of the river coursing between its banks. Some of us are in a hurry to get across. We need to make it for an appointment. Time is money, and we cannot afford to waste it. So to save time, we get into the car and take the road bridge. As everyone else is doing the same, we end up stuck in traffic and fume with impatience. We just cannot reach our destination quickly enough! The motorcyclist, however, does not go fast to save time. He goes fast because it makes him feel more alive. And to escape into life is equivalent to jumping the bridge and re-entering the current. Like the currents of our river, torrential in midstream but sluggish in the shallows, currents of life can also run fast and slow. Whether fast or slow, however, life breaches the connections between points as the river current breaches the lines between its banks. And it leaves us with a choice. We can join with the speeding motorcyclist, going fast in the midstream, shooting the rapids, inhabiting the swerve. Or we can go slow, and hug the banks, exploring its intricate backwaters by foot and wheel. There's a certain freedom to be found in both. But either way, we'll have to take leave of our seats and regain the saddle.

Marginality: From Theory to Practices

Rossella Moscarelli

Abstract The essay tackles the concept of regional marginality, by presenting the main open fields of discussion regarding: the criteria by which to identify the marginal regions, how to measure marginality and the nature of the dynamics between center and periphery. Focusing on these three aspects, the paper concentrates on the relation between theories and consequent practices. Indeed, it is demonstrated how starting from different ideas it is possible to arrive at completely different hypothesis of actions. The paper aims at providing a critical reading about answers given to these complex questions. What makes a region marginal in respect to another? How to measure and represent this phenomenon? Can the unbalanced relationship between core and periphery change in time? If yes, should public policies try to rebalance such relationship? In which way? In the conclusion, an approach to face the issue of marginality is proposed, in the form of "line-based projects", in opposition to the traditional "point-based projects".

1 Marginality as a Concept

Defining marginality is more or less a "mission impossible". "Marginality", such as its synonym "peripherality", is an extremely fuzzy concept: lacking clarity and difficult to test, measure and localize (Markusen 1999). This essay attempts to clarify some aspects of this term, the ones most useful to go from theory to practices.

The starting point is the only theoretical concept usually shared: the dual nature of marginality and peripherality in relation with a center or a core, namely the fact that marginality is always relative to the core (Domansky and Lung 2009). Indeed, being marginal implies a status or a process in which something or somebody is pushed into a secondary position compared to something or somebody. Periphery and marginal are both the opposite and the antithesis of what is center or core. Being peripheral and marginal relates to something that is central, according to a specification of center or close, which indicates that a comparison of the two terms could be done through

R. Moscarelli (✉)
Politecnico di Milano, Milan, Italy
e-mail: rossella.moscarelli@polimi.it

P. Pileri and R. Moscarelli (eds.), *Cycling & Walking for Regional Development*, Research for Development, https://doi.org/10.1007/978-3-030-44003-9_3

substantiating their common opposite. It is not by chance that all the definitions of something peripheral or marginal tend to be the opposite of the focus of the attention. Alternatively, the areas which do not have characteristics or features of core may define the periphery. Indeed, whereas the centers are characterized for what they have and for their potentiality, peripheral and marginal places are defined because of what they are deprived of. Marginal areas are usually defined for what they do not have and for their distance (remoteness) from the places that have the centers. Sthör (1982) argued that peripheral areas are places of low accessibility, regarding the access to markets, to production factors, to private and public services and to sources of social innovation and of economic and political power. According to Ferrau and Lopes (2004) the condition of marginality can be described in terms of distance (to the center), dependence (on the center) and difference (from the center). In short, without a center, there is no periphery.

The academic origins of interest in the dichotomy core and periphery can be traced to the late Fifties, in the economic field, with Perroux (1955), Myrdal (1957) and Hirschmann (1958) and then in the Sixties with the works of Alonso (1964) and Averitt (1968). Actually, already in the Thirties, the geographer and economist Christaller (1933/1966) elaborated the theory of central places that was rediscovered in the Sixties when the English edition was published. This literature shows how peripherality is produced by the mechanism of exploitation of the periphery by the core. Both cores and peripheries are continuously being produced by different coexisting mechanisms. Some of these processes are external (exploitation and dominance), while others are internal to the areas.

In time, after these first studies, several scholars coped this topic, discussing mainly (1) the criteria for defining and identifying marginality, (2) the way to measure marginality and (3) the nature of the relationships between core and periphery.

In the next paragraphs, we will consider the debate regarding these points. They are in part interrelated, but to be clear we will consider them separately.

2 A Criterion to Define Marginality: Lack of Proximity

The criteria by which we can define what is marginal with respect to a core are clearly related to the causes generating condition of marginality.

Herrschel (2012) points out that the condition of peripherality is associated with a distance decay. This means that the remoteness, or in other words the lack of proximity, is one of the most common criterion by which marginality is measured.

The remoteness was usually intended as geographical distance (Fig. 1) from a presumed core/center, i.e. remoteness or edgeness—that is, being at the edge of an area in relation to its center, often expressed by indices of peripherality (Copus 1999; Schürmann and Talaat 2000).

Such original spatial meaning of the term peripherality/marginality, associated with the economic and social penalties faced by locations at a physical distance from

Fig. 1 Physical peripherality. In a spatial sense, is generally a boundary or outer part of any space or body (periphery at the edge) or the space buffering two cores (periphery in between). In the figure, core is the black square and periphery is the white one. Source: elaboration of the author

the main hubs, has been joined by a range of analogous meanings, which have to do with socio-economic marginality in an a-spatial sense (Copus 2001).

In this respect, according to Bock, "whereas in the past, the main cause [of uneven development] was ascribed to geography, this has changed in the sense that the lack of resources is now explained as resulting from a lack of socioeconomic and political connections ('connectivity') and, hence of relational 'remoteness' that is not necessarily bounded to geographical location… Geographical remoteness, as such, therefore does not cause marginalization, nor does central location promise prosperity". (Bock 2016: 556).

In sum, physical or geographical closeness is only one dimension of proximity. This concept was deeply studied by a fringe of economic geographers that since approximately 20 years ago have developed the so-called "proximity school" (Rallet and Torre 1999; Lagendijk and Oinas 2005; Knoben and Oerlemans 2006). In their studies, they have elaborated a theoretical framework to understand the relation between economic activities and the territorial dynamics. Probably the most important tenet of the proximity school is the introduction of the five dimensions of the proximity. In addition to the geographical dimension, Ron Boschma (2005) has identified four others: cognitive, social, institutional and organizational. These dimensions can be intended in this manner:

"Cognitive proximity refers to the extent to which two actors share the same knowledge base (Nooteboom 1999). Social proximity is generally associated with personal relationships between actors (Uzzi 1996), e.g. resulting from past collaboration (Breschi and Lissoni 2009). Institutional proximity is high when actors operate under the same set of norms and incentives, e.g. when co-located in the same country (Gertlerm 1995; Hoekman et al. 2009), or operating in the same social subsystem in particular within academia, industry or government (Etzkowitz and Leydesdorff 2000; Ponds et al. 2007). Finally, organizational proximity refers to the membership to the same organizational entity, as is the case, for example, for two subsidiaries of the same parent company (Balland 2012)." (Balland et al. 2015: 909).

This kind of complex relations was also defined as organized proximity, which indicates the level of social, legal and institutional connection (Torre and Rallet 2005).

2.1 How to Measure Marginality

What we have discussed above, the criteria to classify marginality, concerns the factors that generate conditions of marginality. These can be defined as drivers of marginalization. As Pelc (2018) points out, the issue concerns not only the nature of the drivers, but also the way to measure them, the threshold being the crucial problem. Indeed, in the research of geographical marginality, the indicators of marginality that could be used to evaluate the marginality point—the point from which the individuals, social groups or territories can be defined as marginal, have not yet been widely accepted. Déry et al. (2012) raised this question too, referring to Gino Germani (1980) who accepted that there may be some partial marginality, and the above authors went even a step further to "point to the fact that marginality is always partial, somewhere in-between, since no one can be fully marginal or fully not-marginal … marginality is always 'grey'." (Déry et al. 2012: 8). They consider this dilemma as "similar to the one around the urban-rural dichotomy (and of threshold values in general): where do towns and villages begin?" (ibid.) (Table 1).

Moreover, another issue correlated with the measuring of marginality is the multi-scalability of the concept itself, which can be seen from the urban to the regional scale. Indeed, even if we fix the criteria and a threshold to measure and thus to identify marginality, if we change the scale of measure, we will obtain completely different maps. And, in the same way, if we change the unit of evaluation, again we will obtain different results. These two points, the criteria and methodology by which to measure marginality, are relevant concerning the approach of interventions on peripheral areas. Each policy or strategy, in the past and now, proposes its point of view about the concept of marginality, explicitly or not. We can clarify it by taking as examples two recent strategies about the European Inner Peripheries (Espon 2017) and the Italian Strategy for Inner Areas, SNAI the Italian acronym, (Uval 2014). They propose different perspectives about the idea of peripherality and a different measuring method (Table 2). Obviously, the resulting maps of marginal areas are different (Fig. 2). This is not a secondary point, since these maps identify areas where to allocate special funds and policies.

On the one hand, the European Report defines as "Inner Peripheries" those regions considered peripheral not only because of their geographical remoteness, but also

Table 1 Comparison among different ideas of marginality: inner peripheries (**a**) and inner areas (**b**)

	Criteria to define marginality	Scale to measure marginality	Unit of evaluation
(a) Inner Peripheries (Espon 2017)	Spatial and a-spatial remoteness from main hubs	European	Province (NUT 3)
(b) Inner Areas (UVAL 2014)	Spatial remoteness from main hubs	National (Italy)	Municipality

Table 2 Definitions of personal/regional inequality and social/territorial exclusion

Personal inequality	Regional inequality	Social exclusion	Territorial exclusion
The uneven distribution of income or other variables across different people (Piketty 2014)	The uneven distribution of income or other variables across different locations (adaptation from the definition of social exclusion by Piketty 2014)	"The access for all citizens to basic resources, social services, the labour market and the rights to participate fully in economic, social and cultural life and to enjoy a standard of living and well being that is considered normal in the society in which they live" (EU 2004)	It marks not only differences in the quality of the space and its links to social differences but it also means deprivation in terms of access to a basic degree of urban life and opportunities (Rolnik 2001)

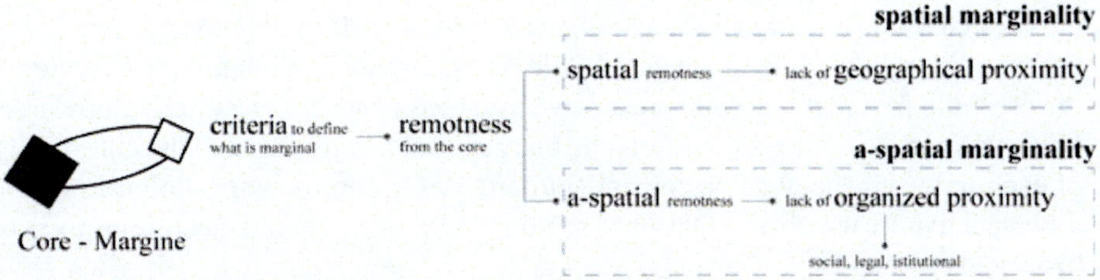

Fig. 2 The figure shows the difference between spatial and a-spatial marginality. *Source* elaboration of the author

because of the lack or low level of proximity to the main social, political and economic activities (Copus et al. 2017). The drivers used to classify inner peripheries are four: poor economic potentials and poor socio-economic situation; lack of access to centers and/or services; poor accessibility and poor economic potentials/poor socio-economic situations. The mapping was done on a European scale and the unit of evaluation is the NUT 3, namely the provinces.

On the other hand, the Italian strategy classifies "Inner Areas" as those territories, in particular those municipalities, substantially far from municipalities classified as "centers" offering essential services, such as education, mobility and healthcare. Thus, in this case, marginality is identified as geographic marginality, through the criterion of spatial remoteness. For what concerns the methodology, the mapping of marginality was carried out on a national scale, using the municipality as unit of evaluation.

These two different approaches demonstrate that there is an open question. What makes a region marginal? The concept lacks precision in at least two senses: in the sense of what constitutes marginality (the criteria), and in drawing the boundary of such an area, even if the meaning itself could be clearly defined (the methodology).

2.2 *The Core-Periphery Dynamic*

In the general debate about the marginality, the studies on the dynamics between core and periphery are some of the most interesting and challenging. They try to answer this question: could the actual marginal places become centers in the future? Or, in other words, is the condition of margin/periphery a stable status or a temporal condition?

Actually, the words peripheralization or marginalization suggests a process, an action that creates a state of being peripheral or marginal in relation to a core. This concept implies the idea that marginality is not a stable status, but rather a process of creation, reproduction and/or breaking out of peripherality. Domanky and Lung explain the idea well:

> the position of the periphery is not determined once and for all. The peripheral status of countries may be subject to change if they manage to achieve success in enhancing their position and change their relations to the core. (Domanky and Lung 2009: 9).

The economic historian Sidney Pollard, in his book Marginal Europe: The Contribution of Marginal Lands since the Middle Ages, provides an interesting overview of the essential role that marginal areas have assumed in several developing processes. In his volume, he demonstrates how the margin could, at times, be at the cutting edge of economic and social advancement (Pollard 1997). In this sense, Pollard suggests a perspective in which the marginal regions have been centers according to some historical events.

3 Should We Care About Regional Marginality?

If we assume that marginality is not a stable status, then there is another point to deal with, about the causes generating marginality and that determine what is core and what is periphery. Speaking of which, Herrschel points out:

> Peripherality may also be actively created – whether intentionally or not. This may occur in the form of exclusions and marginalization of actors through the ways in which policies and power are implemented and defined. (Herrschel 2012: 40)

In this sense, public policies, intended as each decision that Governments take or not (Dye 1987), have a central role in generating or stemming regional marginality. Even in this case, theory and practice are intertwined. At the basis of such line of reasoning, we have to keep in mind another question. Why should the public actor be worried about the uneven development between different places? We do not take for granted the answer, since it is not so predictable.

As we have seen above, the concept itself of marginality arises from an unbalanced relationship between two elements: core and periphery. Such relation, in turn generated by uneven development, can be seen as a coin with two faces: inequality and

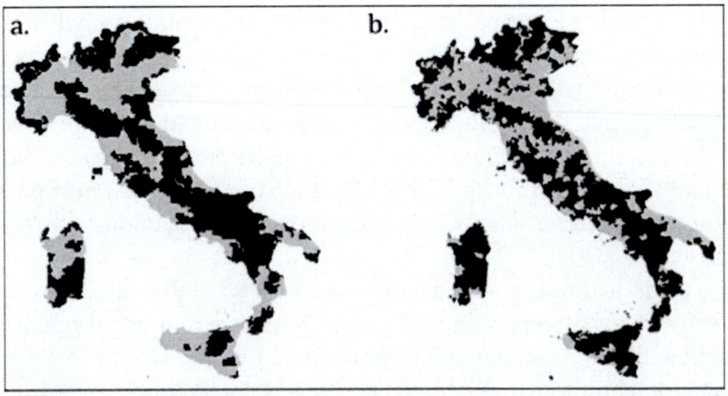

Fig. 3 Comparison among different maps of Italian marginality: extract from the map of European inner peripheries **a** and map of inner areas **b** Elaboration of the author from (**a**) the Map 4.6 "Combining the four delineations: main drivers of inner peripherality" (Espon 2017) and **c** dataset of Italian inner areas

exclusion. Both inequality and exclusion are complex phenomena that can be associated with people and places. In the former case, we will speak about *personal inequality* and *social exclusion.* In the latter, *regional inequality* and *territorial exclusion* (Fig. 3).

Table 2 summarizes useful definitions of these terms, as we will intend in this paper.

At this point, we reach a crossroad on the dilemma about what a policy should consider, whether regional or personal inequality. Citing Rosés and Wolf's words (2018), it seems easy and obvious to answer that "*after all, we should care about people, not about regions as such*".

In the regional economies' field, there are two opposite approaches to this issue. (1) The space-neutral or spatially blind policies, that consider the personal inequalities and the social exclusion. (2) The place-based policies, that consider both the personal and territorial inequalities and the social and regional exclusion. The former is designed without explicit consideration of space. The latter are tailored for the specific characteristics of the territory they are going to be applied.

Two highly influential reports about regional development policy intervention support, respectively, the space-neutral (Word Bank 2009) and the place-based (Barca 2009) approaches. They are inspired by different sources, adopt different philosophies and reach what seems to be very different policy conclusions. The main difference has been centered on the fundamental question of whether efficiency should be concentrated in the core (space-neutral) or there is potential for growth and development in every territory (place-based) (Barca et al. 2012).

The space-neutral approach, also called "spatially-blind", is firmly grounded in three pillars. The first is the new economic geography theory, which advocates the advantages associated with the agglomeration effects of large cities (Duranton 2015): development and growth are inevitably unbalanced and attempt to spread economic activity will not only reduce poverty, they will also undermine growth and prosperity. The second is the Rostovian vision of economic history, which predicts that today's emerging countries will follow the past trajectories of developed states (Mallick 2005). There is, thus, a path that all territories must tread and it involves a greater concentration of economic activity in dynamic poles (Rostow 1962). The third is the conventional spatial equilibrium view of regional economies, with its assumption of perfectly competitive labour, mobile factors of production and mobile labour (Partridge et al. 2015).

The result is a development model based on spatially blind strategies as the most effective way of generating efficiency, guaranteeing equal opportunities and improving the lives of individuals wherever they live and work (World Bank 2009).

For those defending spatially blind approaches, intervention regardless of context is the best way to resolve the dilemma (explained above regarding the personal and territorial inequality) of whether development should be about people or about places. Advocates of spatially blind approaches underscore that "agglomeration and encouraging people mobility not only allows individuals to live where they expect to be better off, but also boosts individual incomes, productivity, knowledge, and aggregate growth." (World Bank 2009: 77). From this perspective, spatially blind policies are also seen as people-based policies, since, firstly, they aim at delivering an improvement in people's lives, regardless of where they live. It is assumed that, by encouraging mobility, spatially blind strategies ultimately lead to a more even geographical distribution of wealth and to the convergence of lagging areas. Consequently, development intervention should be space-neutral and factors simply encouraged to move to where they are most productive, primarily in cities, as the best option to improve individual lives and foster overall economic growth (Gill 2010).

Rodriguez-Pose (2018) has strongly criticized this idea, saying that one of the problematic points of the spatially blind approach lay in the assumption in urban economics that mobility is costless, overestimating the capacity and willingness of individuals to move. He writes:

"Those that remain are unlikely to relocate for a variety of reasons. These include, among others, emotional attachment to the place where they live, age, and lack of sufficient skills and qualifications. Older and retired citizens are much less likely to migrate. Those with limited or conventional skills will be unable to compete at the top end of the London market with the pool of workers with higher skills and, at the bottom end, with those, often coming from abroad and in irregular situations, willing to accept lower wages. The assumption of perfect mobility is, hence, less than perfect." (Rodriguez-Pose 2018: 201).

On the other hand, the idea that the opposite position, the place-based approach, restricts mobility by encouraging people not to move out of places is not true. As Barca explains (2009: 37), "place-based policies are, in fact, intended to enhance individuals' substantive freedom of deciding whether to stay (and to make the most of staying) or to move (and to make the most of moving)." (Barca 2009: 37).

This position is in line with the assumption that each place has its potential for growth and development. Thus, the public policies should promote the full development of each different territories by tailoring place-based strategies. As Barca argues:

"A place-based policy is a long-term strategy aimed at tackling persistent underutilization of potential and reducing persistent social exclusion in specific places through external interventions and multilevel governance." (Barca 2009: VII).

The place-based approach assumes that geographical context really matters, whereby context here is understood in terms of its social, cultural and institutional characteristics. Not only the space matters, but it also shapes the potential for development both territories and, through externalities, of the individuals who live in them. The well being of individuals is influenced by the wider context of the place in which a person happens to live. From this perspective, agglomeration, even if regarded as one of the key drivers of growth, is not seen as the unique model of territorial development. Especially in the case of Europe, where space for large agglomerations to grow is limited and polycentrism is high, economies of scale and growth can be generated by networking between major agglomerations and their hinterland and by dense networks of big or middle-sized cities (Iammarino et al. 2017).

4 Yes, We Should Care

Coming back to the previous question (should we care about regional marginality?), it is clear that these two policies approaches will respond in different ways. Yes, we should care is the answer of place-based approach. There is more than one reason for which it is reasonable that public policies care about regional marginality. We will take into account just some of them, starting from the fact that situations of spatial peripherality can also negatively affect personal conditions. From this perspective, Fernández and Riveiro (2018) argue that there are risks in considering just the people inequalities, since territorial exclusion can imply also social exclusion. Indeed, people that do not have characteristics that can bring them to be socially excluded, such as a low level of income or scholar, are excluded just because they live in excluded territories. In this sense, territorial exclusion has not to be regarded as the localization of social exclusion. Rather, it identifies those territories suffering from exclusion and, consequently, the people living there. In addition, Rosés and Wolf (2018) have discussed the relation between personal and territorial inequality. They show that the trend of territorial inequality appears similar to the trend of personal inequality: there seems to be a growing territorial divergence, with a long-run trend that has recently intensified. Some regions are growing dynamically, while others are stagnating and falling behind. This is evidence that the processes of marginalization are affecting some regions, those that are being left behind. They conclude affirming that such regional divergence can be a problem since "the rapid decay of formerly prosperous parts of Europe can undermine social peace and political stability. In those regions, it is not only the poor who lose but the middle-income groups that are faced with a choice between migration (at the risk of losing their identity) and staying (at the risk of further decline). Or, put differently, between 'exit' and 'voice'. Given our new evidence, we see that this is part of a very broad and long-run change, and very likely driven by the same forces that account for the rise in personal income inequality. Both personal and regional inequality need to be addressed, or else we will observe more of a populist backlash." (Rosés and Wolf 2018)

These reflections suggest that there is a double level of risk in leaving some places behind. The first one concerns what we will abandon in marginalized areas. Indeed, shrinkage and depopulation, characterizing marginal regions, beyond affecting the state of social inclusion of their inhabitants as we have seen above, can cause the loss of essential services (e.g. primary schools in the case of places with a low rate of births), but also historical and cultural, material and immaterial heritage (Di Biase 2013). The second level of risk concerns how the regional marginalization can affect the entire country and those places regarded as centers. This appears in the environmental sphere, since marginal regions, that usually include rural, hilly and mountains areas, are essential for the maintenance and management of the natural and hydrological systems of the entire country (Uval 2014; Borghi 2017). In the political sphere, since people feel abandoned in these areas, due to a sense of "revenge", they can undermine the national socio-political stability and supports populist movements (Rodriguez-Pose 2018). Moreover, we have to remember that exploiting unused potential in intermediate and lagging areas is not detrimental to aggregate growth, but rather, can actually enhance both growth at a local and a national level (Farole et al. 2011). Raising the potential of noncore outlying areas in a synchronized manner may enhance national growth by a significant amount (Garcilazo et al. 2010).

In sum, in some ways, it could be more interesting to consider that core and periphery as communicating vessels, namely as two systems in contact. As Pollard (1997) affirms in his "complementary thesis", margins are neither necessarily inferior nor superior, but are different and often fulfil a complementary role to the regions located in the "core". Complementarity implies a degree of mutual dependence rather than inferiority or superiority.

5 How to Face the Issue of Marginality

If we decide to consider regional marginality as an issue to care about, then we need to decide *how* to face it. This *how*, namely which approach and strategy, depends once again on the idea that we have of marginality. In some way, we need to come back to the criteria we have used to classify peripherality because different ways to see marginality and proximity, as explained above, contribute in elaborating different hypotheses to regenerate shrinking territories (Fig. 4).

On the one hand, if marginality is caused by a geographical distance the actions proposed will aim at increasing what we will call *hard infrastructure,* namely by:

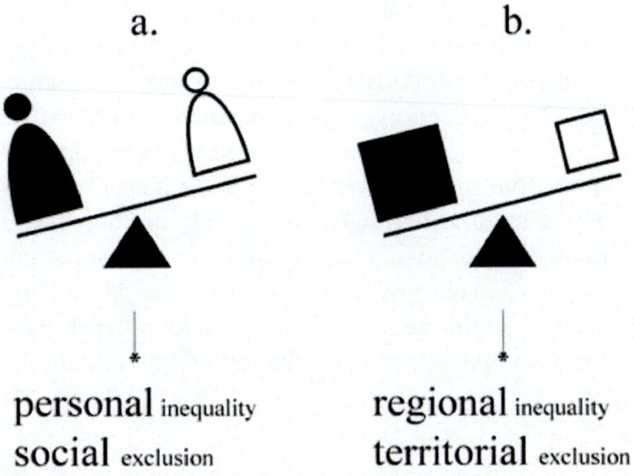

Fig. 4 The figure shows the unbalanced relationship between people (a.) and places (b.) and what it produces. *Source* elaboration of the author

(1) building new infrastructure (roads and facilities) to decrease travel time between core and periphery, improving reachability;
(2) exploiting the recent ICT technologies able to reduce the importance of the geographical distance itself.

If the problem is geographical distance, the difficulty in reaching core areas, the solution will be providing better physical infrastructures.

On the other hand, if we consider also other kinds of distances (or we could say other kinds of lack of proximity), the usual hard infrastructure is not enough. We should elaborate intervention also regarding *soft infrastructure*, related to human resources in wider sense (Makarychev 2004).

Furthermore, if peripheries are "linked to disconnectedness and thus marginalization from the main ("standard") developments by their inconvenient position" (Makarychev 2004: 300), they are simply seen as laggards, which need to be encouraged in catching up with core areas (Herrschel 2012). This implies imagining development strategies, usually effective in core areas, to also regenerate marginalized regions. On the contrary, if peripheral areas are seen through their peculiarities, namely for *what they have* and *what they are*, it is possible to foster strategies specifically suitable for them, different from the models applied in other contexts. Such peculiarities can represent the territorial capital, defined as "the stock of assets which form the basis for endogenous development in each city and region, as well as [...] the institutions, modes of decision-making and professional skills to make best use of those assets" (OECD 2001: 13). Just to provide an example, in the Italian Strategy for Inner Areas, territorial capital is listed as the sum of: (1) environmental capital (water resources, agricultural systems, forests), (2) cultural capital (archaeological assets, historical settlements, churches and abbeys, museums) and (3) cognitive/human capital (landscape, craftsmanship, know-how in traditional activities and productions, local foods, habits and dialects, social energy of the local population) (Uval 2014).

Some scholars (Fratesi and Perucca 2017; Mazzola et al. 2018; Romão and Neuts 2017) maintain that the territorial capital can trigger an alternative model of development, with a consequent regeneration for marginal areas. Moreover, there is an emerging reflection on the experience economy (Toffler 1970; Scott 1997, 2006) as a way to reactivate more isolated areas (Lorentzen 2012). Since in this economic model the most relevant thing is the experience of the resources of a place, such as the cultural heritage, the natural environment, the local tradition or the typical know-how, the place itself matters. In experience economy, place is integrated in the product as a narrative, and it can be tasted as a distinctive experience connected to the service. In this sense, geographic and cultural distance, otherwise seen as a barrier for development, can even be turned into attractions.

6 Line-Based Projects for Marginal Areas…and More

Deciding to deal with territorial marginality is making a stand. We have seen this in every passage of this essay: from when we discuss what is marginal and how it can be measured, to when we are divided between those who consider it necessary to deal with the increase in territorial, as well as social, imbalances, and those who do not. Presenting the different factions, it is now up to us to take a position and decide which side we are on. The proposal of this essay, but also underlying the whole volume, is among those who feel that for various reasons—economic, environmental, social, political and cultural—dealing with the territorial rebalancing is an absolute urgency. How to deal with it? In order to answer this, it is necessary to introduce hypotheses, choices themselves and not absolute and indisputable truths, which guide the formulation of a proposal.. The first hypothesis is that centers and inner areas have to be seen as communicating vessels, namely as systems influencing each other. From this perspective, projects that are activated and realized along a line, that we will call line-based, can be able to define new paradigms and relationships between core and peripheries by imaging mixed strategies (neither just for centers, nor just for peripheries).

The choice of exploiting the lines as designing tools is born from the second hypothesis at the base of such proposal. It regards the fact that the local development strategies are usually elaborated and applied through what we will call a point-based approach. Since point-based projects do not proceed along lines, they carry the risk of fostering the administrative fragmentation and the lack of cooperative vision. On the contrary, line-based projects, by generating an organic vision, starting from the line and fostering a cooperative model, can improve the level of proximity (spatial and also a-spatial) between places. This point can become clearer if we better define what we mean by line-based project. Or better still, with projects that move along lines, of one kind or another. We are considering those projects that are able to hold together the places along a natural line (a river, a valley, a mountain system), or an anthropic line (an abandoned railway, a pilgrimage path, a cultural or historical connection). A sample of line-based projects, or projects that proceed along lines,

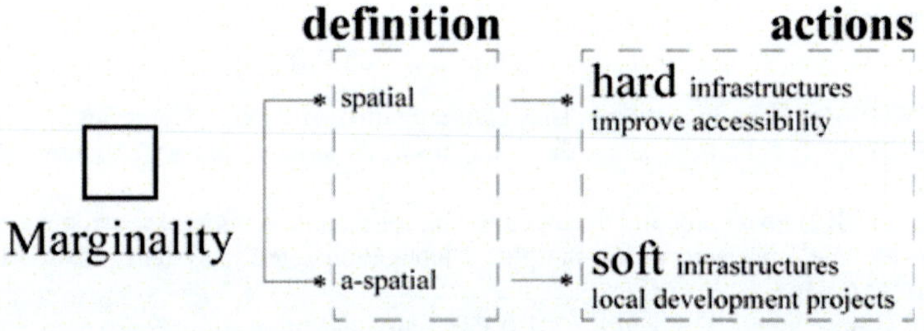

Fig. 5 The figure shows the relation between definitions of marginality and regenerative approaches. *Source* elaboration of the author

are those infrastructures able to activate a slow movement, we can call it "travel", in the territory, such as pathways, cycle paths, bridleways, historical railways, etc. The length and, thus the boundaries, of this kind of project is defined by the length of the line from which the project takes its identity and meaning (e.g. the length of the river, of the valley, of the historical path, etc.). Using this criterion, the sense of the relationship between places, generated by a line, goes beyond the local fragmentation of the administrative boundaries. To recover this relationship, a long line, as long, at least, as long a river, a valley or along path. From this perspective, a path, if it is developed and intended as a territorial project in its complexity, can implies a construction of a cooperative vision and actions both between local actors (horizontal) (Fig. 5) and local and over local actors (vertical) (Dematteis 2001).

Indeed, the places aligned along this infrastructure, whether these areas are marginal or not, are forced to dialogue with each other, to align their policies in some way, to collaborate and integrate their skills, knowledge and resources. This aspect is fundamental if we consider that one of the main reasons for marginalization is isolation, distance, not only geographical, but also political and in relation to the centers (Fig. 6).

Line-based projects can be suitable to tackle the issue of marginalization not only because they redefine the relations between core and peripheries and they foster the cooperation between places and administrative actors, but also since they give new values to the peculiarities of each territory by holding them together along a line and by making them recognizable in a new identity, one related to the line itself.

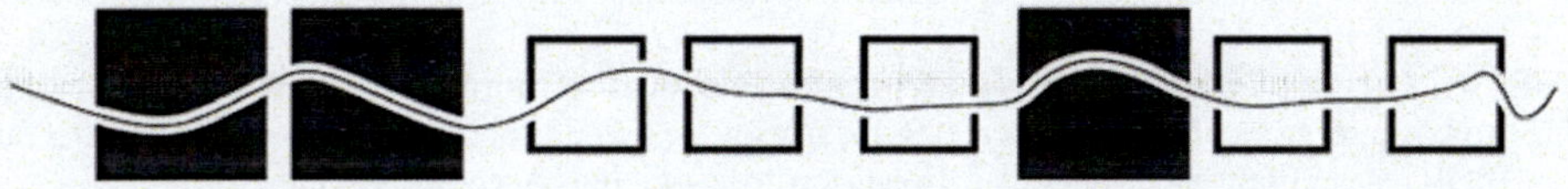

Fig. 6 The image shows the concept of horizontal cooperation along a project of line. In the figure core areas are the black squares and marginal areas are the white ones. *Source* elaboration of the author

References

Alonso W (1964) Location and Land Use. Harvard University Press, Cambridge

Averitt R (1968) The dual economy: the dynamics of American industry structure. W. W, Norton, New York

Balland PA (2012) Proximity and the evolution of collaboration networks: evidence from research and development projects within the global navigation satellite system (GNSS) industry. Reg Stud 46:741–756

Balland PA, Boschma R, Frenken K (2015) Proximity and Innovation: From Statics to Dynamics. Reg Stud 49(6):907–920

Barca F (2009) An Agenda for A Reformed Cohesion Policy: A Place- Based Approach to Meeting European Union Challenges and Expectations. Independent Report, Prepared at the Request of the European Commissioner for Regional Policy. European Commission, Brussels

Barca F, McCann P, Rodríguez-Pose A (2012) The Case for Regional Development Intervention: Place-Based Versus Place-Neutral Approaches. Journal of Regional Science 52(1):134–152

Bock B (2016) Rural Marginalization and the role of Social Innovation: a turn towards nexogenous development and rural reconnection. Sociologica Ruralis 56(4):552–573

Borghi E (2017) Piccole Italie. Le aree interne e la questione territoriale, Donzelli Editore, Roma

Boschma RA (2005) Proximity and innovation: a critical assessment. Reg Stud 39:61–74

Breschi S, lissoni F (2009) Mobility of skilled workers and co-invention networks: an anatomy of localized knowledge flows. J Econom Geogr 9:439–468

Christaller W (1966) Central places in Southern Germany. Translated from the 1933 German edition by C. Baskin. Englewood Cliffs, NJ: Prentice-Hall

Copus A (1999) A new peripherality index for the NUT III regions of the European Union. https://urbanauapp.org/wp-content/uploads/Urbana-Autumn-2001-Volume-VI-Number-1-Andrew-K.-Copus.pdf. Accessed 15 Sep 2019

Copus A (2001) From core-periphery to polycentric development: concepts of spatial and aspatial peripherality. Eur Plan Stud 9(4):539–52

Copus A, Mantino F, Noguera J (2017) Inner Peripheries: an oxymoron on real challenge for territorial cohesion? Italian Journal of Planning Practice, VI I(1):24–49

Dematteis G (2001) Introduzione. Tema, articolazione e risultati della ricerca. In: Dematteis G, Governa F (eds) Contesti locali e grandi infrastrutture. Politiche e progetti in Italia e in Europa. Franco Angeli, Milano

Déry S, Leimgruber W, Zsilincsar W (2012) Understanding marginality: Recent insights from a geographical perspective. Hrvatski geografski glasnik 74(1):5–18

Di Biase C (2013) Small towns in inland areas: shared landscapes. In: Crisan R, Fiorani D, Kealy L et al (eds) Conservation – Reconstruction. Small historic centers conservation in the midst of change, EAAE, Hasselt, pp 131–139

Domansky B, Lung Y (2009) Editorial: the changing face of the European periphery in the automotive industry. European Urban and Regional Studies 16:5–10

Dye TR (1987) Understanding public policy, III edn. Englewood Cliffs, NJ, Prentice Hall

Duranton G (2015) Growing through cities in developing countries. World Bank Research Observer 30:39–73

Espon (2017) PROFECY – Inner Peripheries: National territories facing challenges of access to basic services of general interest. https://www.espon.eu/programme/projects/espon-2020/applied- research/inner-peripheries-national-territories-facing. Accessed 15 Sep 2019

EU European Commission (2004) Joint report on social inclusion, https://ec.europa.eu/employment_social/social_inclusion/docs/final_joint_inclusion_report_2003_en.pdf. Accessed 15 Sep 2019

Etzkowitz H, Leydesdorff L (2000) The dynamics of innovation: from national systems and'Mode 2'to a triple helix of university–industry–government relations. Res Policy 29:109–123

Farole T, Rodrıguez-Pose A, Storper M (2011) Cohesion Policy in the European Union: Growth, Geography, Institutions. JCMS: Journal of Common Market Studies, 49(5):1089–1111

Fernández M, Riveiro D (2018) A exclusion territorial como unha forma de manifestación dos procesos de exclusión social. SÉMATA, Ciencias Sociais e Humanidades 30:145–165
Ferrau J, Lopes R (2004) Understanding peripheral rural areas as contexts for economic development. In: Labrianidis L (ed) The Future of Europe's Rural Peripheries. Ashgate, Aldershot
Fratesi U, Perucca G (2017) Territorial capital and the resilience of European regions. Ann Reg Sci 60(2):241–264
Garcilazo JE, Martins JO, Tompson W (2010) Why Policies May Need to be Place-Based in Order to be People-Centred. OECD Regional Development Policy Division, Paris
Germani G (1980) Marginality. Transaction Books, New Brunswick
Gertlerm S (1995) 'Being there': proximity, organization, and culture in the development and adoption of advanced manu-facturing technologies, Econom Geogr 71:1–26
Gill I (2010) Regional Development Policies: Place-Based or People- Centred?. OECD Regional Development Policy Division, Paris
Herrschel T (2012) Regionalisation and marginalization: bridging old and new divisions in regional governance. In: Danson M, De Souza P (eds) Regional development in Northern Europe: peripherality, marginality and border issues. Routledge, London, pp 30–48
Hirschman A (1958) The strategy of Economic Development. Yale University Press, New Haven
Hoekman J, Frenken K, Vanoort F (2009) The geography of collaborative knowledge production in Europe. Ann Reg Sci 43:721–738
Iammarino S, Rodríguez-Pose A, Storper M (2017) Why Regional Development matters for Europe's Economic Future. Working papers of European Commission
Knoben J, Oerlemans L (2006) Proximity and inter-organizational collaboration: a literature review. International Journal of Management Reviews 8:71–89
Lagendijk A, Oinas P (2005) Proximity, external relations, and local economic development. In: Lagendijk A, Oinas P (eds) Proximity, Distance and Diversity. Issues on Economic Interaction and Local Development, Ashgate, Aldershot, pp 3–22
Lorentzen A (2012) The development of the periphery in the experience economy. In: Danson M, De Souza P (eds) Regional development in Northern Europe: peripherality, marginality and border issues. Routledge, London, pp 16–29
Makarychev A (2004) Where the North meets the East: Europe's "dimensionalism" and Poland "marginality strategy". Cooperation and Conflict: Journal of the Nordic International Studies Association 39(3):299–315
Mallick OB (2005) Rostow's Five Stage Model of Development and its Relevance to Globalisation. University of Newcastle, Newcastle
Markusen A (1999) Fuzzy Concepts, Scanty Evidence, Policy Distance: The Case for Rigour and Policy Relevance in Critical Regional Studies. Reg Stud 33(9):869–884
Mazzola F, Lo Cascio I, Epifanio R et al (2018) Territorial capital and growth over the Great Recession: a local analysis for Italy. Ann Reg Sci 60(2):411–441
Myrdal G (1957) Economic Theory and Under-developed Regions. Gerald Duckworth, London
OECD (2001) Territorial Outlook 2001. Organization for Economic Cooperation and Development, Paris
Partridge MD, Rickman DS, Olfert MR et al (2015) When spatial equilibrium fails: is placebased policy second best? In: McCann P, Varga A (eds) Place-based economic development and the new EU Cohesion Policy. Routledge, New York, pp 49–71
Pelc S (2018) Drivers of Marginalization from Different Perspectives. In: Pelc S, Koderman M (eds) Nature, Tourism and Ethnicity as Drivers of (De)Marginalization. Springer International Publishing, Insights to Marginality from Perspective of Sustainability and Development, pp 3–27
Perroux F (1955) Note sur la notion de pôle de croissance. Économie appliquée 8:307–320
Piketty T (2014) Capital in the 21st century. Harvard University Press, Boston
Pollard S (1997) Marginal Europe: the contribution of marginal lands since the Middle Ages. Oxford University Press, Oxford
Ponds R, Vanoort FG, Frenken K (2010) Innovation, spillovers and university–industry collaboration: an extended knowledge production function approach. J Econom Geogr 10:231–255

Rallet A, Torre A (1999) Is geographical proximity necessary in the innovation networks in the era of global economy? GeoJournal 49:373–380

Rodríguez-Pose A (2018) The revenge of the places that don't matter (and what to do about it). Cambridge Journal of Regions, Economy and Society 11(1):189–209

Rolnik R (2001) Territorial exclusion and violence: the case of the state of São Paulo. Geoforum Brazil 32(4):471–482

Romão J, Neuts B (2017) Territorial capital, smart tourism specialization and sustainable regional development: Experiences from Europe. Habitat International 68:64–74

Rosés J, Wolf N (2018) The return of regional inequality: Europe from 1900 to today. CEPR Policy Portal. https://voxeu.org/article/return- regional-inequality-europe-1900-today. Accessed 15 Sep 2019

Rostow WW (1962) The Stages of Economic Growth. Cambridge University Press, London

Schürmann C, Talaat A (2000) Towards a European Peripherality Index. Final Report, Report for General Directorate XVI Regional Policy of the European Commission, https://ec.europa.eu/regional_policy/sources/docgener/studies/pdf/pe riph.pdf. Accessed 15 Sep 2019

Scott AJ (1997) The cultural economy of cities. Int J Urban Reg Res 21(2):323–339

Scott AJ (2006) Creative cities: conceptual issues and policy questions. Journal of Urban Affairs 28(1):1–17

Stöhr WB (1982) Structural characteristics of peripheral areas: the relevance of the stock-in-trade variables of regional science. Papers of the Regional Science Association 49(1):69–84

Toffler A (1970) Future Shock. Bantam books, New York

Torre A, Rallet A (2005) Proximity and localization. Reg Stud 39:47–59

Uval (2014) A strategy for Inner Areas in Italy: definition, objectives, tools and governance, Issue 31

UZZI B (1996) The sources and consequences of embeddedness for the economic performance of organizations: the network effect. Am Sociol Rev 61:674–698

World Bank (2009) World Development Report 2009: Reshaping Economic Geography. World Bank, Washington

Marginalised Areas as a Public Policy Concern

Manos Matsaganis

Abstract This chapter is an attempt to sketch the broader context into which regional development for 'left behind' places, and the potential for their regeneration via 'slow tourism', the main focus of this book, is located. This context does not lend itself to easy generalisations. This becomes clear in Section One, which summarises recent changes in the distribution of incomes—at personal, regional and country level—in Europe and beyond, and finds that while inequality between countries has fallen, inequality within countries (whether between rich and poor families, or between dynamic and lagging or declining areas) has often increased. Section Two offers a brief account of the drivers of income inequality, focusing on the effects of trade (globalisation) versus technology (automation) on jobs and earnings, and emphasising local effects. Section Three reviews the current debate on the political consequences of regional decline, as manifest in electoral support for populist forces, in the light of competing paradigms in economic geography ('place-based' versus 'people-based' policy approaches). The final section discusses the implications for public policy for the revitalisation of lagging and declining areas, stressing the need to steer a new course, tapping into unused potential and local knowledge, and involving all relevant stakeholders.

1 Have (Regional) Inequalities Risen in Recent Decades?

Inequality is "the defining issue of our time," as US President Barack Obama famously declared in 2013. But inequality between what exactly? The widely shared assumption that inequality is on the rise rests on the fact that the distance between rich and poor has grown considerably within countries (for instance, in the US and the UK). But the opposite is true with respect to inequality between countries: this has fallen in recent decades, whether we compare China and India to the West, or the Central and Eastern European countries to older EU members. A similar process of between-country convergence versus within-country divergence manifests itself

M. Matsaganis (✉)
Polytecnic University of Milan, Milan, Italy
e-mail: emmanuel.matsaganis@polimi.it

P. Pileri and R. Moscarelli (eds.), *Cycling & Walking for Regional Development*, Research for Development, https://doi.org/10.1007/978-3-030-44003-9_4

at sub-national level: regions in poorer countries have covered some of the distance that separated them from those in richer ones. But this image of convergence, that holds when transnational entities such as the EU or the OECD are taken as a whole, breaks down when the gaze turns to regional inequalities within the same country: in Italy, the UK, the US and elsewhere, these have grown in recent years.

Specifically, if we pitch our analysis of inequality at global level, that is, if we treat all individuals in the world as members of the same community, irrespective of where they live, and consider their incomes over recent decades, in terms of purchasing power (i.e. adjusted for differences in the cost of living), we will conclude that the global income distribution has actually become less unequal. The decline in global inequality (from 0.74 in 1990 to 0.69 in 2010, as measured by the Gini coefficient) was "almost exclusively a result of the decline in inequality between countries, with the average inequality within countries remaining more or less constant until the turn of the millennium but increasing afterward" (Bourguignon 2015, p. 43).

The visual image of changes in the global income distribution is the 'elephant curve' (Milanovic 2016). Income gains since the late 1980s were significant among some (but not all) of the world's poor, and spectacular around the middle of the world income distribution. The persons involved (90% of which lived in China, India and other Asian economies such as Thailand, Vietnam and Indonesia) were the greatest beneficiaries of globalisation, and formed the "emerging global middle class." In contrast, higher up the world income distribution, income growth was virtually zero. The persons belonging to that group tended to be found in the bottom half of the income distribution of the country where they lived, which overwhelmingly was the US, or Japan, or Germany, or some other advanced economy. Globalisation has clearly not benefited the "lower middle class in the rich world." On the other hand, the 'global plutocracy' (about half of which were US citizens) saw their fortunes soaring: the richest 10% of the world income distribution captured about 60% of the total gains in real incomes between 1988 and 2008; the top 1% alone accounted for as much as 19% of all income gains over the period.

The rapid industrialisation and fast growth of a few developing countries, coupled with the deindustrialisation and low growth of developed countries, brought about a 'Great Convergence' (Baldwin 2016). As recently as 1990, the seven largest advanced economies (the US, Germany, Japan, France, Britain, Canada and Italy), known collectively as 'the G7', produced 65% of world manufacturing; China only 3%. By 2010, China's share had grown to 19%, while that of the G7 had shrunk to 47%. The rise of China, India and other emerging economies caused the global income distribution to rebalance: the share of the G7 in all world income declined from two-thirds in 1990 to approximately 45% in 2014, which happened to be about the same as in 1900. accelerated as a result of the global financial crisis and its aftermath. The turmoil in stock exchanges and falling property prices dented the incomes of the 'global plutocracy', while low or no growth in the advanced economies caused those of the "lower middle class in the rich world" to stagnate. Meanwhile, even though China as a whole remained less prosperous than the poorest member states of the EU, the distance was fast disappearing: average incomes in urban areas of China doubled in 2008–2011, exceeding (in purchasing power terms) those in Romania, Bulgaria, Latvia or Lithuania.

The emergence of a 'global plutocracy' is probably the most glaring manifestation of a broader phenomenon: namely, the recent rise of inequality in almost all advanced industrialised countries. The longest time series available, pieced together by the late Tony Atkinson and his colleagues in The Chartbook of Economic Inequality (Atkinson et al. 2017), show that the Gini coefficient declined continuously in Italy, from 0.49 in 1901 to 0.29 in 1991, before it rose again to 0.35 in 2012 (the same level as in the mid-1970s). As for the share of the richest 1% in total gross income (excluding capital gains), it went from 6.3% in 1983 to 9.9% in 2007. Italy hardly stood out among advanced economies in this respect: the share of the top 1% rose much higher in the US (18.9% in 2012), and to a lesser extent in Britain (15.4% in 2009) and in Germany (13.9% in 2008), while it was lower in France (9.3% in 2007) and in Sweden (7.2% in 2013).

Rich and poor are not randomly distributed between countries, nor between regions within the same country. Have regional inequalities risen or fallen in recent years? Again, it depends on the level of analysis. If we consider each region in a group of countries separately, then it is likely to conclude that regional inequalities have declined. At the same time, inequalities between regions within the same country have in many cases increased. A recent World Bank report showed that inequality between EU member states, as measured by the coefficient of variation in GDP per capita, declined by almost 30% in 2000–2009, and then fluctuated around more or less that level. On the other hand, inequality between EU regions also declined in 2000–2009, though less dramatically, and increased thereafter, being as high in 2016 as it was in 2005. As for inequality between regions within the same EU member state, it rose rather steadily over the period, most steeply in 2004–2010, and was 12% higher in 2015 than it was in 2000 (Farole et al. 2018, p. 36). Similarly, an OECD report found that:

> When considered at the scale of the OECD area, economic inequality across regions declined since the turn of the millennium. Between 2000 and 2015, inequality in regional GDP declined by 15% across the OECD and by 25% across Europe, driven by the catching up of regions in countries with comparatively lower income. However, in many OECD countries the economic disparities between the regions within the same country grew substantially over the same time period. [...] As of 2015, per capita GDP levels in almost two-fifths of OECD regions (135 out of 350) were still below their pre-crisis peak. (OECD 2018a, p. 13).

In Italy and Spain in particular, "the negative correlation between growth and initial level of per capita GDP—indicating the catching up of less affluent regions to the more prosperous ones—turns positive for the 2008–15 period. This means that less-developed regions in these countries were not only unable to narrow the gap, but they even lost ground against more affluent parts of the country." (OECD 2018a, p. 21)

In fact, the standard practice of using per capita income often conceals an even grimmer picture: 'left behind' places, as in Italy's Mezzogiorno, decline demographically as well as economically. The outward migration of unemployed youth may improve statistics, but reinforces the vicious circle of decline (Cannari et al. 2019,

Viesti 2019). At EU level, there seems to be strong correlation between income level and population change: in 2000–2014, the richest European regions (GDP per capita 150% or more of EU average) saw their population rise by nearly 10%; over the same period, the poorest regions (GDP per capita less than 75% of EU average) experienced a 2% fall in population (Iammarino et al. 2019).

2 Recent Drivers of (Regional) Inequality

What are the causes of rising inequality in advanced economies such as the US and Europe? Abstracting from political economy explanations (such as the decline in union membership), the simplest way to begin thinking about the drivers of inequality is to consider the components of household incomes: participation of men and women (and their grown-up children) in the labour market, earnings from employment (and other market activities), household formation, taxation and other public policies (e.g. as regards pensions and other social benefits). Starting from the latter, the decline in tax progressivity, in the form of lower tax rates on wealth, inheritances, companies and high earners, has certainly contributed to the inequality of post-tax incomes. Less obviously, so has the proliferation of single person households and of 'assortative mating', or the tendency of men and women to choose partners from the same socioeconomic group and/or of similar education. The distribution of jobs has also become more polarised: at household level, with two-earner and no-earner (or 'low work intensity') households both being more common; at regional level, in Italy, in 2018, the female employment rate varied from 60.7% in the North-East to 32.7% in the South (the male employment rate was 75.6% and 57.3%, respectively).

Earnings (from work) have also grown unequal in recent decades. For a long time, economists debated the relative contributions to greater earnings inequality of technology versus globalisation (Autor et al. 2015). According to an early insight, "skill-biased technological change" increased the skills premium, widening the distance between the wages of high-skilled workers relative to those of low-skilled ones. A later, more nuanced version established that technology increased wage polarisation: it rendered redundant and/or compressed the earnings of workers specialised in routine tasks, while at the same time raising the demand for both high-skilled jobs at the upper end of the earnings scale, and low-skilled non-routine service jobs at the bottom of the distribution (Autor et al. 2003; Goos and Manning 2007). On the other hand, the disappearance of well-paid jobs in manufacturing in the US and Western Europe, as firms relocated to lower wage countries, or were driven out of business by cheap imports from China and other emerging economies, was also associated with rising inequality. As has been pointed out, the two leading contenders for the recent rise in inequality (technology and globalisation) may in fact offer complementary explanations:

> [T]he lower price of capital goods leading to the replacement of routine labor and greater complementarity between capital and high-skilled workers [...] could have occurred only under the conditions of globalisation, where reduced prices of capital goods were made

possible thanks to the existence of cheap labor in China and the rest of Asia. (Milanovic 2016, pp. 109-110)

The effects of both technology and globalisation are far from uniformly distributed across space. Recent estimates of the share of jobs at high risk of automation (consisting of tasks that are likely to be automated in the near future with a probability of over 70%) found it ranged from 4.3% in the Norwegian region of Oslo and Akershus to 39.3% in West Slovakia. In Italy the risk of automation varied across regions within a narrower range, from 13.7% of jobs in Lazio to 15.6% in Marche. Adopting a broader definition of jobs at 'significant' risk of automation (probability 50%–70%) resulted in variation between 28.8% and 70.0% of jobs across regions (OECD 2019, pp. 85–86). Several regions managed the transition from high- to low-risk jobs in recent years. In Italy, in 12 regions more jobs were lost than created between 2011 and 2016; of the remaining, most net job creation took place in Lombardy and Lazio, followed by Campania. In the first two, the jobs created were predominantly in occupations at high risk of automation; in Campania, the opposite was the case (OECD 2018b, pp. 207–208).

Nevertheless, technology creates some jobs while it destroys others (Acemoglu and Restrepo 2019, Autor 2019, Bessen forthcoming), with the net effect being difficult to predict. So far, automation seems to have had less of an impact on unemployment as such, while it has shifted employment towards non-routine jobs, both at the high- and the low-end of the earnings scale. In contrast, evidence from the US (Autor et al. 2013, 2016), Norway (Balsvik et al. 2015), Germany (Dauth et al. 2014), and Spain (Donoso et al. 2015) suggests that the industries affected by competition from imports (and hence the jobs vulnerable to trade shocks) are likely to be clustered in certain regions. Summing up, "[f]oreign competition can force firms out of business and can lead to the disappearance of entire industries in extreme cases," while in contrast automation "leads to job losses among workers whose jobs get automatized, but does not destroy the underlying fabric of firms within a region" (OECD 2018a, pp. 129–131).

3 Economic Geography and "The Revenge of the Places that Don't Matter"

Should public policy be concerned with inequalities? The answer is less self-evident than it may seem. The libertarian tradition of Hayek (1944) and Friedman (1962) has long denied the need to redistribute income and fight social inequalities (except perhaps to ensure that no one in society starves), on the grounds that attempting to do so is both futile and harmful. There are echoes of that in the debate on regional inequalities, but opposition to, or scepticism for, regional development policies tends to be more subtle.

At a purely theoretical level, from the perspective of the neoclassical growth paradigm "if market forces were allowed to function properly and without hindrance," persistent regional divergence should not exist at all.

> In a world without places-pecific scale economies where the goods produced as well as factors of production (or even ideas) can move freely across space […] it is hard to imagine why, say, London would flourish and the north of England would not; or why growth and incomes in Italy's Mezzogiorno would lag for any substantial period behind those of the so-called Third Italy or Northern Italy. Faced with the stylised facts that clearly show that regional disparities do exist and persist, under a neoclassical perspective the justification for policy intervention is simply to 'free up' the flow of labour and capital to regions where their return is the highest. That is to say, labour should move to labour-scarce regions and capital should flow to capital-scarce regions. With a properly working price mechanism, higher wages (in the booming region) would for instance automatically take care of this adjustment process. Facilitating this regional adjustment process could be seen as the main or even only task of regional policy, but one could also argue that what is really called for is a national policy that stimulates factor mobility and price and wage flexibility. (Garretsen et al. 2013, pp. 180–181).

The traditional response of economists critical of the neoclassical paradigm has been to point out that (i) there is nothing self-correcting about regional inequalities, and (ii) regional inequalities are actually bad for national growth. On the first point, Myrdal (1957) argued that, due to a process of "circular and cumulative causation," "in the absence of counteracting policies inequalities would tend to increase, both internationally and within a country" (Myrdal 1984, pp. 152-153). On his part, Hirschman (1958), while rather more relaxed than most development experts at that time about unbalanced growth in general, was careful to stress that "in regional development the process of unbalanced growth is fundamentally different from unbalanced growth in the sectoral sense because of the weakness of the forces making for restoration of interregional balance" (Hirschman 1984, p. 105). Nor did Myrdal share the idea, prevalent at his time, that "increasing internal inequalities were […] an inevitable result of growth." On the contrary, he saw "greater equality as a condition for more substantial growth" (1984, p. 153). In a similar vein, Kaldor (1970) advanced an efficiency argument for regional policy, stressing congestion costs and other 'diseconomies' from excessive growth in some areas (Wishlade 2019, p. 11).

The notion that uneven regional development may actually be the result of market forces is one of the insights of urban economics and the 'new economic geography', stressing the importance of transportation costs for industrial clusters (Krugman 1991), and the advantages for successful cities stemming from agglomeration economies, i.e. "the benefits that come when firms and people locate near one another together" (Glaeser 2010, p. 1). These theoretical advances certainly led to a better understanding of what makes some places successful, and of the path-dependent nature of economic development. However, coupled with a growing realisation that the results of large-scale efforts to industrialise less-developed regions on the part of government agencies like the Cassa per il Mezzogiorno in Italy had been meagre, the policy implication of the new paradigm was twofold: (i) policies to support economic activity in lagging and declining areas should be scaled down, and (ii) those living there should be given assistance to move to cities and other successful areas where employment opportunities were more plentiful. Specifically:

> Focusing on cities and, especially, on the largest and most prosperous cities implies that 'economic growth will be unbalanced' (World Bank 2009, p. xxi), but 'to try to spread it out is to discourage it – to fight prosperity, not poverty' (World Bank 2009, back cover). Hence, the main aim of policies should be to move people to places where there are opportunities, not opportunities to declining areas (World Bank 2009, p. 241). 'Subsidizing poor or unproductive places is an imperfect way of transferring resources to poor people' (Kline and Moretti 2014, p. 656). In sum, this implies pursuing what has been called people-based policies at the expense of place-based policies, as place-based policies are, at best inefficient and, at worst, distortionary and slow the need for structural reform. (Rodríguez-Pose 2018, pp. 191-192)

Of course, greater mobility to successful cities brings with it a new set of 'negative externalities', e.g. in housing markets. To make housing more affordable, while preserving density, cities ought to be allowed to grow, perferrably in height, but also in space. But the real negative externality was likely to be of a different kind:

> [T]elling people that they and their children had no future if they stayed put [amounts to] being patronised by an aloof and distant elite. An elite that was telling them that the alternative was between facing slow decline – possibly attenuated by attempts at decentralising powers – or leaving the places where most of them had been born and bred. An elite that, moreover, would suffer none of the costs and reap the potential benefits of any change in policy. Being told that where you live and where you belong does not matter is bound to make anyone react. And reacted they have. The reaction has, however, come from a somewhat unexpected source. In recent years the places that "don't matter" have increasingly used the ballot box (and, in some cases, outright revolt) to rebel against the feeling of being left behind; against the feeling of lacking opportunities and future prospects. (Rodríguez-Pose 2018, p. 190)

In this narrative, rising support for populist forces around Europe and beyond, from the Brexit referendum and the US presidential election in 2016, to a series of electoral contests, in the Netherlands (2017), France (2017), Germany (2018), Italy (2018), Argentina (2019) and the UK (2019), ought to be understood as the "revenge of the places that don't matter" (Los et al. 2017; Essletzbichler et al. 2018; Viesti 2018; Wishlade 2019). A detailed geographical analysis of voting patterns in 2013–2018 in all 28 EU member states confirms that "the anti-EU vote is mainly driven by a combination of long-term economic and industrial decline, low levels of education, and a lack of local employment opportunities" (Dijkstra et al. 2018).

4 Dealing with the Geography of Discontent

The implications for public policy are not straightforward. Lagging and declining regions in Italy and other advanced economies find themselves in a 'middle income trap'. A high road to growth, based on the production of high-quality and high-value goods and services, appears to lie beyond their reach. At the same time, a low-cost strategy would get them nowhere, as they will always be out-competed by the latest emerging economy. Their predicament is that "they are neither as productive nor as innovative as the high- and very-high-income regions, but their labour and land prices are higher than those in low-income regions" (Iammarino et al. 2019). How can this impasse be overcome? Policy solutions will inevitably vary according to the specific combination of challenges and constraints faced by each area. For instance, in "slow-growing regions of southern Europe, [beset by] poor-quality government, pervasive corruption, collusion and lack of trust, [...] institutional improvement is crucial to improving the incentives for skilling and entrepreneurship." In contrast:

> Some ageing and declining industrial areas in France, northern Italy and northern Spain have good-quality government but suffer from skill-mismatch to the current economy, as well as degraded amenity and residential conditions, combined with deepening social problems due to the despair attendant on long-term decline. [...] In such regions, fraying social capital needs to be repaired, through re-creation of networks of workers, government, universities, entrepreneurs and investors to overcome the mistrust [...] inherited from the period of their initial industrialisation (Iammarino et al. 2019).

A place-sensitive strategy for both types of areas, lagging or declining, would avoid the pitfalls of the traditional schools of thought, assuming neither that "market convergence happens naturally as a spatially blind process as long as market barriers are removed," nor that "'administered convergence' is something that can only be achieved via redistribution." Instead, public policy would adopt a "developmentalist approach aiming at maximizing the development potential of all regions" (Barca et al. 2012, p. 146; see also McCann and Ortega-Argilés 2015; OECD 2018a).

What the preceding analysis implies is that policymakers can ignore the plight of marginalised areas at their peril, that it is unrealistic to expect (and unwise to encourage) those living there to abandon them and move to the nearest thriving city, and that a development strategy, to be successful, ought to tap into unused potential and local knowledge. But the success of such a strategy is anything but guaranteed. Policymakers can hardly afford to rely on market incentives alone: that would merely accelerate the trend towards ever-growing polarisation between dynamic and lagging or declining areas. Nor can they afford to fall back on past practices: they will have to invent innovative approaches anew, involving local stakeholders, building trust among participants, allowing for flexibility and public debate, but having first clearly defined the policy objectives to pursue, the responsibilities of all participants, and the 'conditionalities' for the provision of public support (Barca et al. 2012). This is likely to prove a tall order, but nothing less may do.

References

Acemoglu D, Restrepo P (2019) Automation and new tasks: how technology displaces and reinstates labor. J Econ Perspect 33(2):3–30

Atkinson T, Hasell J, Morelli S, Roser M (2017) The chartbook of economic inequality. INET Oxford Working Paper no. 2017–10. Institute for New Economic Thinking at the Oxford Martin School

Autor D (2019) Work of the past, work of the future. Am Econ Assoc Papers Proc 109:1–32

Autor D, Levy F, Murnane R (2003) The skill content of recent technological change: an empirical exploration. Q J Econ 118(4):1279–1333
Autor D, Dorn D, Hanson G (2013) The China syndrome: local labor market effects of import competition in the United States. Am Econ Rev 103(6):2121–2168
Autor D, Dorn D, Hanson G (2015) Untangling trade and technology: evidence from local labour markets. Econ J 125(584):621–646
Autor D, Dorn D, Hanson G (2016) The China shock: learning from labor-market adjustment to large changes in trade. Ann Rev Econ 8:205–240
Baldwin R (2016) The Great Convergence: information technology and the new globalization. Harvard University Press, Cambridge, MA
Balsvik R, Jensen S, Salvanes K (2015) Made in China, sold in Norway: local labor market effects of an import shock. J Public Econ 127:137–144
Barca F, McCann P, Rodríguez-Pose A (2012) The case for regional development intervention: place-based versus place-neutral approaches. J Regional Sci 52:134–152
Bessen JE (forthcoming) Automation and jobs: when technology boosts employment. Econ Policy
Bourguignon F (2015) The globalization of inequality. Princeton University Press, Princeton
Cannari L, Viesti G, Zanardi A (2019) Interregional disparities in Italy: structural changes and public policies, a brief introduction. Politica Econ/J Econ Policy 2:159–172
Dauth W, Findeisen S, Suedekum J (2014) The rise of the East and the Far East: German labor markets and trade integration. J Eur Econ Assoc 12(6):1643–1675
Dijkstra L, Poelman H, Rodríguez-Pose A (2018) The geography of EU discontent. Directorate-General for Regional and Urban Policy, Working Paper 12/2018. Luxembourg: Publications Office of the European Union
Donoso V, Martin V, Minondo A (2015) Do differences in the exposure to Chinese imports lead to differences in local labour market outcomes? an analysis for Spanish provinces. Reg Stud 49(10):1746–1764
Essletzbichler J, Disslbacher F, Moser M (2018) The victims of neoliberal globalisation and the rise of the populist vote: a comparative analysis of three recent electoral decisions. Cambridge J Regions, Econ Soc 11:73–94
Farole T, Goga S, Ionescu-Heroiu M (2018) Rethinking lagging regions: using cohesion policy to deliver on the potential of Europe's regions. World Bank, Washington, DC
Friedman M (1962) Capitalism and freedom. Chicago University Press, Chicago
Garretsen H, McCann P, Martin R, Tyler P (2013) The future of regional policy. Cambridge J Regions, Econ Soc 6(2):179–186
Glaeser E (2010) Introduction. In: Glaeser E (ed) Agglomeration economics. Chicago University Press, Chicago
Goos M, Manning A (2007) Lousy and lovely jobs: the rising polarization of work in Britain. Rev Econ Stat 89(1):118–133
Hayek FA (1944) The road to serfdom. Routledge, London
Hirschman AO (1958) The strategy of economic development. Yale University Press, New Haven
Hirschman AO (1984) A dissenter's confession: the strategy of economic development revisited. In: Meier GM, Seers D (eds) Pioneers in development. Oxford University Press, Oxford
Iammarino S, Rodriguez-Pose A, Storper M (2019) Regional inequality in Europe: evidence, theory and policy implications. J Econ Geogr 19(2):273–298
Kaldor N (1970) The case for regional policies. Scottish J Politic Econ 17(3):337–348
Kline P, Moretti E (2014) People, places, and public policy: some simple welfare economics of local economic development programs. Ann Rev Econ 6:629–662
Krugman P (1991) Increasing returns and economic geography. J Polit Econ 99(3):483–499
Los B, McCann P, Springford J, Thissen M (2017) The mismatch between local voting and the local economic consequences of Brexit. Reg Stud 51(1):786–799
McCann P, Ortega-Argilés R (2015) Smart specialization, regional growth and applications to European Union cohesion policy. Reg Stud 49(8):1291–1302

Milanovic B (2016) Global inequality: a new approach for the age of globalization. Harvard University Press, Cambridge, MA

Myrdal G (1957) Economic theory and underdeveloped regions. Gerald Duckworth, London

Myrdal G (1984) International inequality and foreign aid in retrospect. In: Meier GM, Seers D (eds) Pioneers in development. Oxford University Press, Oxford

OECD (2018a) *Productivity and jobs in a globalised world: (how) can all regions benefit?*. Organisation for Economic Co-operation and Development, Paris

OECD (2018b) *Job creation and local economic development 2018: preparing for the future of work*. Organisation for Economic Co-operation and Development, Paris

OECD (2019) *Regional outlook 2019: leveraging megatrends for cities and rural areas*. Organisation for Economic Co-operation and Development, Paris

Rodríguez-Pose A (2018) The revenge of the places that don't matter (and what to do about it). Cambridge J Regions, Econ Soc 11:189–209

Viesti G (2018) La vendetta delle regioni che non contano. Il Mulino 3:451–458

Viesti G (2019) Qualche riflessione sulla nuova geografia economica europea. Meridiana 94:137–164

Wishlade F (2019) The rise of populism, regional disparities and the regional policy response. European Policy Research Paper No. 109. Glasgow: University of Strathclyde

World Bank (2009) World Development Report 2009: reshaping economic geography. Washington, DC: World Bank

Italian Policies on Marginal Territories: An Overview

Benedetta Silva

Abstract The topic of inner areas and attention to economically marginal areas manifest periodically. The paper traces the main public policies that deal with these areas since the second post-war period. The literature review intends to critically evaluate the results that these interventions have given to the territories. The study of the first results of the National Strategy for the Inner Areas deals with the theme of participated planning and the bottom-up experience in regional cohesion policies: the areas selected become an experimental laboratory and an incubator for solutions identified *ad hoc*. The paper, therefore, focuses on the effects of the Intervento Straordinario per il Mezzogiorno (Extraordinary experience for South Italy) and of the actions for the economically depressed territory of the Centre-North. The research identifies differences, overlaps and problems that led to the suppression of the two measures: for 40 years these interventions have attempted to reform the agricultural sector and try to favour the establishment and development of the industrial sector. The critical reading focuses therefore on the analysis of the specific legislation for mountain areas: in this case, the legislation from a first phase of a welfare nature changes towards a cultural policy of protection and enhancement of local particularities, losing its specificity in time.

1 The National Strategy for Inner Areas as example of Regional Cohesion Policies

National Strategy for the Inner Areas has a particular importance in the 2014–2020 program for regional cohesion policies. It represents a *unicum* at European level: it is a strategy that coordinates rural development policies and cohesion policies in a multi-sectoral project for the development of rural areas.

B. Silva (✉)
Department of Architecture and Urban Studies, Politecnico di Milano, Milan, Italy
e-mail: benedetta.silva@polimi.it

P. Pileri and R. Moscarelli (eds.), *Cycling & Walking for Regional Development*, Research for Development, https://doi.org/10.1007/978-3-030-44003-9_5

The theme of policies for inland areas has an ancient tradition in Italy: there is no single definition because the "inner area" concept is constantly evolving (Mantino 2015).

The Department for development policies, starting from a polycentric interpretation of the national territory, has developed a Map of Inner Areas between 2012 and 2014 (Strategia Nazionale per le Aree Interne: definizione, obiettivi, strumenti e governance 2014). This classification overcomes the historical distinction between city and countryside, between mountain and plain; it defines the territories spatially distant (in travel times) from the centres offering citizenship services (identified in health, education and mobility). The Inner Areas represent a "national question": these involve more than 60% of the national territory and 52% of the Municipalities, including about a quarter of the population (Fig. 1). "Inner Area" is not necessarily synonymous of "weak area", but it identifies the degree of peripherality, referring exclusively to the services considered (Barca et al. 2014). These are areas that, despite having undergone a process of gradual marginalization after the Second World War and being significantly distant from the centres offering services, are rich in environmental and cultural resources and are characterized by a strong potential: local development actions can concern the protection of the territory and natural and cultural resources, agro-food systems and sustainable tourism, local crafts, energy savings and renewable energy resources.

The Italian Government and the European Commission, through the Partnership Agreement, have decided to concentrate interventions in the territories classified simultaneously as Inner Areas and rural areas type C and D[1] of the Rural Development Plan (Accordo di partenariato 2017). Italian "Leggi di Stabilità" (financial laws), with national funds, have provided for a total commitment of 281.18 million euros for the adaptation of services since 2014, to which must be added the allocation of 50 million by the INAIL real estate investments (funds for the construction of schools only), while local development actions are financed with all available EU funds, FERS, FSE, FEASR e FEAMP[2] (Relazione annuale sulla strategia per le Aree Interne 2018).

By creating a Technical Committee, the Italian Government together with the Regions has identified a limited number of "project areas": since 2015, through in situ visits and forums with local territorial actors, they evaluated the critical points and strengths of the different territories. In 2016, 68 areas were identified, to which another 4 areas were added the following year, for a total of 72 areas (1071 Municipalities). These represent an experimental laboratory to test the flexibility of the legislation developed, a learning by doing system that much differs from the rigidity of previous policies. The Strategy, by prescribing the need to work in partnership, has undoubtedly favoured relations between local actors and represented an important boost to the associative processes of the Municipalities involved, favouring the stipulation of agreements to manage local public services (need denounced also in

[1]This classification defines the eligible areas of the PSR programm: type C indicates the intermediate rural areas in decline, while type D the rural areas with development problems.

[2]It is not yet possible to quantify this funding because not all project areas have completed the process for signing the Area Strategies.

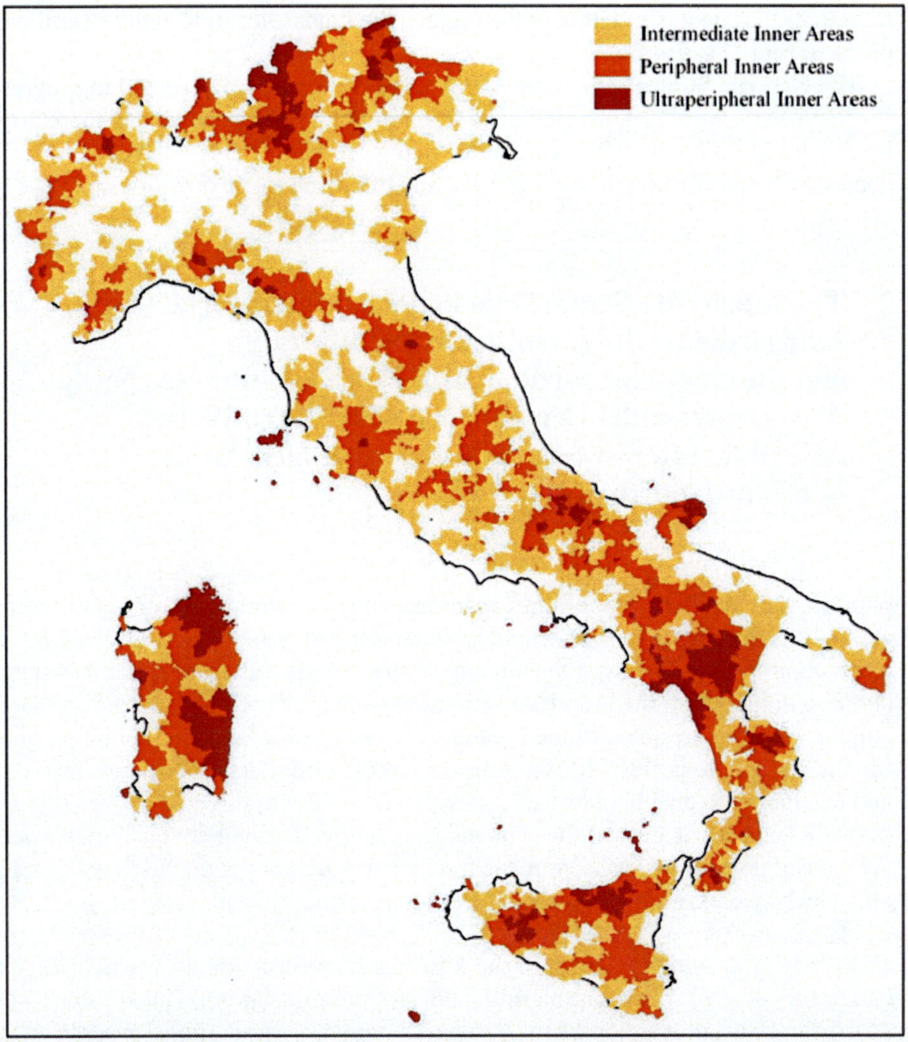

Fig. 1 Map of Inner Areas, 2012. Intermediate Inner Areas are defined between 20 and 40 min in travel times from the centres offering citizenship services; Peripheral Inner Areas between 40 and 75 min, while Ultraperipheral areas overcome 75 min (*Source* old2018.agenziacoesione.gov.it, elaboration by the author)

Romano and Fiorini 2018). Through participatory planning involving the population, stakeholders and different levels of administration (Municipalities adhering to the partnership, Regions, central government and European Commission), the Strategy represents an extraordinary incubator of solutions: each territory, investigating the critical issues and local potential, develops an ad hoc designed intervention program.

As of October 2019, 55 Strategy Preliminaries have been approved, of which 38 have already been converted into Area Strategies. The drafting of the Area Strategies represents the conclusive phase of the project, the funds allocation and the fulfilment of the territorial actions.

The National Strategy for Inner Areas is the most recent effort, still in progress, to intervene on territorial marginalities. Other policies have responded to these problems in the past. These policies were directed to South of Italy, to the economically depressed areas of Central-North of Italy and to the mountain territories.

2 The Intervento Straordinario per il Mezzogiorno (The Extraordinary Intervention for South Italy) and the Opere Straordinarie di Pubblico Interesse per le Aree Depresse del Centro-Nord (The Extra Works for Public Interest for the Underdeveloped Areas of Central-North of Italy)

In the aftermath of the end of the Second World War, Italy appeared to be overpopulated, dramatically poor in raw materials and in conditions of strong economic and political dependence. The whole national territory was strongly affected by the destruction of the war. Given the already severe economic situation, the Resistance chose to do not harm the industrial activities in the north of Italy (Borghi 2018). In August 1944 the president of the United States Roosevelt launched an aid program for the Italian areas freed by the allies, a credit in dollars to buy raw materials and machinery to enable industrial recovery. This aid was directed above all in the northern and central areas of the peninsula, excluding the southern territories, where the industrial sector was weaker: in the immediate post-war period this flow of capital affected the pre-existing economic gap between North and South, increasing it (Galisi 2014).

In 1950, following the begin of the post-war reconstruction, it was necessary to tackle the so-called "questione meridionale" (southern question), (Felice 2007; Galisi 2014): the extra intervention for the South was started, as an additional program to the ordinary State resources, with an inter-sectoral nature and multi-year duration, and the "Cassa per le opere straordinarie di pubblico interesse nell'Italia Meridionale" (Casmez—Cassa del Mezzogiorno).[3] Cassa del Mezzogiorno, initially intended as an institution with decision-making autonomy with respect to political power, found its models in two experiences of the 1930s, the American Tennesee Valley Authority (1933) and in the Italian strategy of IRI—Institute for Industrial Reconstruction (Felice 2007; Lepore 2011): it was intended to operate with interventions that provide for quicker and leaner procedures than those of ordinary administration, with the assignment of loans that exceed the annual schedule and allow for the performance of coordinated activities and not small works distributed over time (Lepore 2011).

[3]Law 10 August 1950, n. 646.

The objectives of the provision were to finance initiatives aimed at the economic development of the South and eliminate, or at least diminish, the gap with the Northern areas. At first, the priority was to modernize the agricultural sector; a second phase would then focus on the location of the industry in Southern areas. Economists understood that it was not possible to overcome the North–South gap with the only increase in the agricultural and tourism sectors. Considering the huge workforce present, it was necessary to invest also in the industrial sector. There was no convenience to invest in Southern areas and due to the lack of accumulation of productive capital. Therefore State intervention was indispensable: these measures were both direct and indirect actions. The firsts acted through the creation of agricultural infrastructures, road and rail communication networks, extension of the system of irrigation and aqueducts for water supply. The seconds aimed to investment incentives, non-repayable grants and low-interest loans for private companies (Galisi 2014).

The territorial jurisdiction of the provision was in the regions of Abruzzo and Molise, Campania, Puglia, Basilicata, Calabria, Sicily and Sardinia, in the provinces of Latina and Frosinone, on the Island of Elba, in the municipalities of the province of Rieti already included in the former district of Cittaducale, in the Municipalities included in the area of the Tronto river reclamation area, as well as the Municipalities of the province of Rome included in the area of the Latina reclamation area[4] (Fig. 2).

In evaluating the results of this important initiative, a distinction should be made between a first phase of general success, from 1950 to 1970, and a second phase in which interventions proved to be largely failure (1971–1992). In the first 20 years, as a matter of fact, a moment of pre-industrialization (interventions for agriculture), one of reorientation of funds in the industrial sector[5] and finally a period in which regional politics were tried to be included within national planning. The intensification of industrial incentives led to the creation of 15 "industrial development areas" in cities with at least 200,000 inhabitants and 22 "industrialization zones" in areas with at least 75,000 inhabitants. These incentives were intended for the development of productive activities and the creation of industrial complexes, which could counteract mass emigration with the formation of permanent jobs and the creation of "development poles". In most cases these initiatives proved to be unsuccessful because the "motor enterprises" installed could not widely activate the industrial sector, remaining as "cathedrals in the desert" (Felice 2007). With the Law n. 717/1965 the autonomy of Cassa per il Mezzogiorno declined, because it was obliged to submit the programs for approval to the Ministry for extraordinary interventions in the South; since the 1970s, also due to the oil crisis and the war between Iran and Iraq, the complementarity with the ordinary administration of the State fell away with welfarism and patronage distortions, and the Regions[6] entered the decision-making process. Only during the 1980s it became clear that, despite the high costs for the public budget,

[4]As indicated in Law 10 August 1950, n. 646, art. 3.

[5]Law n. 634/1957.

[6]Instituted with the Law n. 853/1971.

Fig. 2 Areas subject to Intervento Straordinario per il Mezzogiorno (*Source* Law 10 August 1950, n. 646, art. 3, elaboration by the author)

the results often coincided with the objectives set. A reorganization of the extra intervention for the South was therefore necessary, with the suppression of the Cassa per il Mezzogiorno (1984) and its replacement with the Agency for the promotion and development of the South (Agensud).[7] The new institution, even if it was operating in the traditional territorial context (thus not taking into consideration the economic development that has occurred in the meantime), no longer dealt with the planning of the interventions (now delegated to the participatory planning of the Regions and to the Government) but only to the provision of subsidies and funding (Trezzani 1995).

Simultaneously to the creation of the extra intervention for the South, during the 1950s the implementation of extra works of public interest was also prepared for the territories of northern and central Italy, for the so-called "economically depressed

[7]Instituted with the law n. 64/1986.

locations" (Becchi-Collidà et al. 1989). These localities, initially identified according to the discretion of the Governments in agreement with the Committee of Ministers (1950–1957),[8] were only delimited since 1966 according to three criteria impoverishing of the labour force due to aging or exodus of the residents; per capita income lower than the national average and low productivity levels due to problems in agriculture or industrial activities (Law n. 614/1966). Although less known and studied, these were initiatives for the construction of public utility works for agricultural transformation, reclamation and irrigation, ordinary non-state roads, aqueducts and sewers (Law n. 647/1950, art. 1 subsection 2): if the Municipalities with a population of less than 10,000 inhabitants showed that they could not take on the loans for the realization of the interventions, the State could act with the partially or totally taken charge of the costs (Law n. 635/1957). As regards the benefits to companies, the extra intervention for the depressed areas of the Centre-North provided contributions for agricultural activities, subsidized loans for the construction and expansion of industrial companies and tourism, and tax exemptions for 10 years of direct income taxes for new artisan businesses and small industries that settled in the Municipalities referred to as economically depressed.

Both experiences, the one concerning extra intervention in the South and the one for the economically depressed territories of the Centre-North, are based on a top-down driven model: the central administration, seen the structural difficulties, participates with the construction of infrastructures that promote the localization and expansion of productive activities and encourage industrial initiative. However, the facilities did not have to be in competition: initially, in fact, the intervention for the Centre-North was aimed at small private companies (maximum 100 employees), while that for the South was aimed at large public companies. Only since the 1970s, with the re-modulation of the incentives of the extra intervention for the South, even for small and medium-sized enterprises, experiences become alternatives: the policy for the Centre-North in this phase in fact reduces the effectiveness of the one for the southern regions (Felice 2007). This overlapping of intervention combined with an increasingly rigid European Community policy and the referendum procedure for the cessation of the extra intervention in the South (that had developed in the early 1990s) led to the overcoming of extra legislation. The progressive strengthening of the community institutions and the passage of regional development policies from national to European competences led to the suppression of the Department for extra interventions for Southern Italy and Agensud and the transformation into ordinary intervention for depressed areas throughout the national territory (1993).

[8] Law n. 647/1950 and law n. 635/1957.

3 Laws in Favour of Mountain Territories

Another important branch of measures for Italian disadvantaged areas is represented by public policies for mountain areas: in this case the intention is to focus on the general laws specific to mountain areas.

Since the beginning of the Republic, recognizing also the importance of the experience of partisan resistance, the Constituent Assembly has specified in Article 44 of the Constitution the need to provide specific regulations for mountain areas: it meant recognizing a particular value for these territories as disadvantaged compared to the flat areas, both for the physical-morphological characteristics of the places and the anthropic ones, but above all for the greater costs for living these areas (Losavio and Perniciaro 2017; Borghi 2018).

What is, therefore, the definition of mountain areas? The answer to this question was given by the first law issued specifically for mountain areas, Law n. 991/1952. The provision defined, as a matter of fact, the so-called "montagna legale" (legal mountain, Fig. 3): the census commission classified the Municipalities with respect to the altimetric parameter (having at least 80% of the surface above the 600 m above sea level) and to one of economic nature (average taxable income).

In addition, economic incentives, mortgage concessions and tax breaks were identified for the development of farms and for the improvement of hygiene and accommodation for private homes for the tourism function. Both actions were welfare fund: external financial support to territories still engaged in traditional economic activities and often penalized by physical isolation (Mantino 2015).

This approach was resumed and confirmed in 1971 with the new rules for mountain development: here incentives were provided for the support of the integrated mountain economy and compensations for the inhabitants, recognizing the functions of protection of the territory and the conditions of hardship where they live. In this case, the legislator also tried to involve the local populations in the development choices and in the enhancement of the mountain areas to eliminate the imbalances of an economic and social nature and to defend their soil and nature: with the Law n. 1102/1971, in fact, a new institutional subject is born, the so-called Comunità Montana (Mountain Community), an entity with the task of governing the territory and promoting local development. In a rather ambitious program this institution, delimited as a homogeneous area of territory, should have drawn up a multi-year Plan of Interventions for economic and social development financed with funds allocated by the Regions: in reality the task proved more complex than expected due to the difficulties in coordinating the different programming levels and the results were largely ineffective (Romano and Fiorini 2018).

The provisions issued for the mountain territories in the 1990s represent a change in the legislative activity that became of a "cultural" type (Losavio and Perniciaro 2017): in fact, the Law n. 97/1994 provided for territorial, economic, social and cultural actions (Mantino 2015) for the development of mountain territories, through the safeguarding of environmental qualities and potential. For the first time the theme

Fig. 3 Map of the classification of the "legal mountain" (*Source* Law n. 991/1952, elaboration by the author)

of valorisation was accompanied by that of protection: in this sense, one of the operations envisaged by the standard was the creation of a specific register for mountain products. Compensation actions were planned, instead, to avoid the exodus and to favour the establishment and permanence in the mountain Municipalities with population less than 5000 inhabitants: incentives and prizes were provided for those who transferred the residence or the economic activity in the mountain contexts for at least 10 years, through the competition in the costs of relocating and renovating the buildings. The law also established the "National Mountain Fund" which, with additional character with respect to ordinary and special financing, was intended for the Regions with the task of regulating their resources.

With the 2007 crisis (both institutional and fiscal), State transfers decreased: the National Fund was resized and erased in 2011 (Borghi 2018), as were the government transfers for the management of the Mountain Communities, which were suppressed in almost all the Regions.[9] The Comunità Montane (Mountain Communities) are often replaced by the Unioni di Comuni Montani (Unions of Mountain Municipalities). Only in 2013 the "Supplementary National Fund for the mountain municipalities"[10] was established again and destined to entirely mountain municipalities, and in 2016 the "National Mountain Fund" is reactivated for a period of 3 years.

A new legislative update specific to the mountain territories is currently lacking: in recent years it has often tried to revise the law n. 97/1994, but the various attempts have not been successful. Instead, non-specific laws have been issued, but references to mountain territories can be traced: they are the Strategy for Green Communities (L. n. 221/2015), the Strategy of Inner Areas (which is described in the previous paragraph) and the Law for Small Municipalities, L. n. 158/2017, (for a critical reading of the provision, see Romano and Fiorini 2018). Over the last 70 years, the mountain policy has evolved from a first welfare phase which, recognizing the difficulties of life in these areas, provides special facilities for the population, to a cultural phase where the need to protect specificities and local productions, favouring resettlement: at the same time, however, the mountain legislation loses its specificity and comes to be part of a more general regulations.

4 Remarks on the Main National Policies for Marginal Areas

The rereading of the main national policies for marginal or economically disadvantaged areas highlights some important observations.

Attention to these territories presents alternating phases. The lack of continuity has often made sporadic actions (Mantino 2015) and the results are ineffective. The type of governance, on the other hand, changes from a central institutional design to a local integrated design, while the areas of intervention are restricted from very large territories to spontaneous aggregations of Municipalities.

In particular, the Extraordinary Intervention for the South has proved effectiveness in its first phase: the expansion of the hydraulic and transport infrastructures has favoured economic development and the convergence of incomes, decreasing the economic gap between North and South. Starting from the 70s, on the other hand, the interventions for the South have lost incisiveness due to the relief and patronage implications and the partial overlap with the intervention for the economically depressed territories of the Centre-North.

With reference to the measures for mountain areas, it is shown that the interest in these areas, particularly since the Constitution, has progressively lost its specificity, especially in the last 25 years, relegating mountain policies to general measures for disadvantaged areas.

---

[9] Mountain Communities exist only in Lombardy and Campania.

[10] Law n. 228/2012.

The National Strategy for Inner Areas represents an *unicum* within the regional cohesion policies: an experimental program that intends to reverse current trends by promoting the relations between local actors and the different levels of central administration. As of October 2019, the continuity of the experience is not yet guaranteed by the 2021–2027 program.

The process represents an important opportunity for the territories involved: starting from the needs and resources actually available, solutions for local problems are proposed through bottom-up projects. Collaboration between administrations has allowed to overcome the reciprocal mistrust and to start actions among peers. It created an opportunity to imagine the territory together and no longer in a fragmented way, as often happened in the past. Furthermore, the strengthening of the permanent inter-municipal system has become the tool to manage collectively the development of the whole process (Lucatelli and Monaco 2018).

References

Accordo di partenariato (2014–2020) Italia, Sezioni 3 e 4. https://opencoesione.gov.it/media/uploads/documenti/adp/accordo_di_partenariato_sezioni_3_e_4_2017.pdf. Accessed 9 Nov 2019

Barca F, Casavola P, Lucatelli S (2014) Strategia nazionale per le aree interne: definizione, obiettivi, strumenti e governance, in Materiali UVAL, n. 31 year 2014. https://old2018.agenziacoesione.gov.it/opencms/export/sites/dps/it/documentazione/servizi/materiali_uval/Documenti/MUVAL_31_Aree_interne.pdf. Accessed 9 Nov 2019

Becchi-Collidà A, Ciciotti E, Mela A (1989) La natura del problema. In: Becchi-Collidà A, Cicciotti E, Mela A (eds) Aree interne, tutela del territorio e valorizzazione delle risorse. Franco Angeli, Milano, pp 13–36

Borghi E (2018) Piccole Italie. Le aree interne e la questione territoriale, Donzelli, Roma

Bozza della strategia (2019). https://old2018.agenziacoesione.gov.it/it/arint/Strategie_di_area/Bozze_della_strategia.html#. Accessed 9 Nov 2019

Costituzione della Repubblica Italiana

Felice E (2007) Divari regionali e intervento pubblico. Per una rilettura dello sviluppo in Italia. Società editrice il Mulino, Bolognawe

Galisi R (2014) Ricostruzione e programmazione nell'intervento straordinario per il Mezzogiorno. FrancoAngeli, Milano

Losavio C, Perniciaro G (2017) Progetto di ricerca: "Analisi della normativa inerente ai territori montani". https://www.issirfa.cnr.it/analisi-della-normativa-inerente-ai-territori-montani.html. Accessed 9 Nov 2019.

Law 10/08/ 1950, n. 647, Esecuzione di opere straordinarie di pubblico interesse nell'Italia settentrionale e centrale

Law 25/07/1952, n. 911, Provvedimenti in favore dei territori montani

Law 29/07/1957, n. 635, Disposizioni integrative della legge 10 agosto 1950, n. 647 per l'esecuzione di opere straordinarie di pubblico interesse nell'Italia settentrionale e centrale

Law 22/07/1966, n. 614, Interventi straordinari a favore dei territori depressi dell'Italia settentrionale e centrale

Law 3/12/1971, n. 1102, "Nuove norme per lo sviluppo della montagna"

Law 31/01/1994, n. 97, Nuove disposizioni per le zone montane

Law 24/12/2012, n.228, Disposizioni per la formazione del bilancio annuale e pluriennale dello Stato

Law 28/12/2015, n. 221, Disposizioni in materia ambientale per promuovere misure di green economy e per il contenimento dell'uso eccessivo di risorse naturali

Law 6/10/2017, n. 158, Misure per il sostegno e la valorizzazione dei piccoli comuni, nonché disposizioni per la riqualificazione e il recupero dei centri storici dei medesimi comuni

Lepore A (2011) Cassa per il Mezzogiorno e politiche per lo sviluppo. In: Società Italiana degli Storici Economici (ed) Istituzioni ed economia, atti del convegno di studi. Cacucci Editore, Bari, pp 107–166

Lezzi B (2018) Relazione annuale sulla strategia nazionale per le aree interne. https://old2018.agenziacoesione.gov.it/opencms/export/sites/dps/it/documentazione/Aree_interne/Presentazione/Relazione_CIPE_ARINT_311218.pdf. Accessed 9 Nov 2019

Lucatelli S, Monaco F (2018) La voce dei Sindaci delle Aree Interne. Problemi e prospettiva della Strategia Nazionael. Runnettino Editore, Soveria Mannelli

Mantino F (2015) Da Rossi-Doria a oggi: come e perché cambiano le politiche per le aree interne. In: Meloni (ed) Aree interne e progetti d'area. Rosenberg & Sellier, Torino, pp 264–284

Romano B, Fiorini L (2018) Abbandoni, costi pubblici, dispersione. Alla ricerca di migliori risposte. In: Pileri P, Moscarelli R (ed) Italy is an internal area, Urban Tracks n. 26, pp 66–73

Strategia nazionale per le Aree interne: definizione, obiettivi, strumenti e governance (2014). https://www.programmazioneeconomica.gov.it/wp-content/uploads/2017/02/Accordo-P-Strategia_nazionale_per_le_Aree_interne_definizione_obiettivi_strumenti_e_governance_2014.pdf. Accessed 9 Nov 2019

Trezzani C (1995) Dall'intervento straordinario nel Mezzogiorno all'intervento ordinario nelle aree depresse. In: Bollettino di informazioni costituzionali e parlamentari, n. 1–3, pp 305–317. https://bpr.camera.it/bpr/allegati/show/6138_1118_t. Accessed 9 Nov 2019

Landscape and Heritage as Keys for Slow Travel

Slowness to Discover the Ordinary Italian Landscape

Chiara Visentin

Abstract "The history of walking is the story of each of us", Rebecca Solnit wrote. The ordinary Italian landscape, without ostentatious of magnificence, modest but fascinating as it can be scrutinized in its most intimate aspects, is varied and articulated, beautiful and simple, the best place to feel in its slowness of everyday life (walking, cycling, contemplating). From here telling slowness in the Italian landscape between these quite invisible characteristics is certainly a must. Studying the statistical data about Italy we discover that it's a Nation of territorial inequalities, but if we initially suppose that these are well defined between North and South, then we understand that we are wrong: there are gaps both in the North and in the South, both on the mountains and on the flat lands, both along the sea coasts and internal areas, both in rural and periurban areas. A geography where creation of new opportunities not only improves the economic aspect but also the will and empathy of the communities: the goal could be to intend this territory as a unitary cultural heritage, where culture stands for knowledge and participation; a landscape connected to daily life and therefore not superbly evident but a needful production landscape to look at through a new systemic approach.

1 Meaning of the Places

It can become complex to recognize and bring out the heritage "of minor things" scattered in ordinary Italian landscapes. What are the formulas that help this research? What are the characteristics? What are the parameters? What is this ordinary patrimony invisible to most of us? And why could the slowness that flows on threads like cycle paths and walking paths become a rehabilitating factor for this heritage? Indeed, it can be precisely living these places carefully that can make us discover

C. Visentin (✉)
Research Unit, "Architecture and Production Landscapes", IUAV University of Venice, Venice, Italy
e-mail: cvisentin@community.iuav.it

Urban Planning professor, Scuola di Architettura, Urbanistica, Ingegneria delle Costruzioni Polimi, Mantua, Italy

P. Pileri and R. Moscarelli (eds.), *Cycling & Walking for Regional Development*, Research for Development, https://doi.org/10.1007/978-3-030-44003-9_6

elements or moments that we would never be able to grasp in the fast daily routine, because hidden in the folds of silent landscapes.

Give attention to the issues of territory and its "little" identities does not mean *Localism*, but a strategic element on a theoretical and practical level for redefining the models of evolution and balance of places. In Italy, the local dimension is to be considered the dominant theme: the Nation "of the hundred cities", where the medium-small size of urban environment is the evident and obvious characteristic and the connection with respective surrounding territory. Mostly rural.

The time has come to stop thinking only about the "special" beautiful landscape (recognized as a *brand* in Italy and abroad) and to invest in other more realistic representations closer to us, precisely because of those contexts (such as rural landscapes or built by hydraulic reclamation) fully represent us, have formed us and have evolved with us, but that we forget, and we don't look at every day. They are "facts". In these landscapes, which are continuously lived but constantly forgotten to look at, to which little attention is paid, there is also part of the fragility of the internal areas. The knowledge of the places is the result of a daily life of it, accompanied in the sequence of time, seasons, and history.

Most of all *productive landscapes* to be protected but above all to be known and discovered. In slowness. Today we are faced with the serious problem of the loss of soil and agriculture, connected to the dual phenomenon of abandonment and land consumption. "From the second half of the twentieth century, from 1956 to 2001, the urbanized area has increased by 500%, to grow even today" (ISPRA 2014). This is an accelerated, unreasonable and dangerous transformation that has taken a very dangerous line especially in the last 50 years.

Through the discovery of these ordinary environments, the route to change is to reconnect agriculture and community, city and countryside, local use and global economy, specific and general, usual and exceptional. It is no longer possible to avoid the goal of a coherent care of the territory, of environmental and landscape maintenance, of protection from various types of risks, from the hydrogeological to that of fires and floods. The themes of the forest and the water are necessarily complemented by the analysis of rural land and production of food (Figs. 1 and 2). This cannot be separated from the recognition of the farmer's role as a landscape craftsman, food producer, and as a multi-functional worker both in flat lands and on the highlands. One of the exponents of the Neapolitan Enlightenment, Giuseppe Maria Galanti, wrote that "agriculture is that noble art without which no one would exist" (Galanti 1789). The food chain is indeed the main way in the relationship between man and the earth. Food should regain the right centrality between human activities; and with it soil and fertile land must regain significance, but also the understanding of the process that can be done by knowing the territories and following them (Magnaghi 2010). In their rhythms. In the times of the seasons. In slowness (Hutabarat 2009).

Landscape is now considered one of the key themes for the valorization of European identity (Clementi 2002). The importance of improving the specificity, the meaning of the places, the characterization of the space, deriving from identity, cultural (Michell 2003), and relational elements is affirmed (Turri 2003). In this

Fig. 1 Rural lands in East side of Veneto. Photo by Andrea Morucchio ©

Fig. 2 Rural lands in Po Valley, Emilia-Romagna. Photo by Giuliano Ferrari ©

regard we can talk about *la société paysagiste* as defined by Pierre Donadieu: a specific community that consciously or unconsciously built relationships with the territory in which lived, physical, symbolic, ecological relations (Donadieu 2002). A territory intended as a unitary cultural heritage (Fig. 3). The land is considered a single schedule of historical occasions, no longer made up only by physical signs or

Fig. 3 Fontanili lands near Corte Valle Re, Campegine, Emilia-Romagna. The *fontanili* are phenomena related to the springing of deep water that are induced to rise to the surface. The ones in Campegine are the last examples of low-lying springs in Emilia-Romagna. Photo by Chiara Visentin

by material objects, but considered as the result of many circumstances (Starobinski 1999).

Places, landscapes, architectures that don't have an evident monumentality in themselves, but an extraordinary integration in the contexts: real characterizations of the territories, history, and processes that produced them: "a second Nature that works for civil purposes", so at the end of the eighteenth century Goethe defined the Roman agricultural landscape that he traveled in his *Journey to Italy*: well this quote can consistently refer to most of the Italian *production landscape*, that landscape connected to everyday life and therefore not superbly evident (Fig. 4).

2 Minor Landscapes, Fundamental *Artifacts*, Perception and Slowness

If Jose Saramago in one of his most intense stories reminds us that "the most abundant thing on earth is the landscape…" (Saramago 2010), then we can decide to start discovering our landscapes, knowing them. What is the most intense way to reach a totalizing experience? Cross them. Following the slow rhythms of nature. If travel is knowledge, the landscape is its richest source of information. From a simple aesthetic corollary, landscape thus becomes the ideal tool to understand a place, as a

Fig. 4 Rice fields near Carpi Modena, Emilia-Romagna. The cultivation of rice in Modena area has a tradition started in the 1930s. Photo by Chiara Visentin

more extended territory, this manages to transform and distinguish the traveler from the tourist, the inhabitant from a careless visitor. Time for discovery and interaction is fundamental: the slowness in discerning the places is often the real difference to identify also a traveler from the tourist. "The history of walking is the story of each of us", Rebecca Solnit wrote (Solnit 2001). The usual Italian landscape, modest, ordinary, without ostentatious of beauty, but fascinating because it can be scrutinized in its most intimate aspects, is varied, articulated, a place that incessantly tells and collects its past history. The variety obviously depends primarily on its morphological appearance, and when this variety is not present then everything is taken for granted, not spectacular, not important.

From here telling slowness in the Italian landscape between these quite invisible characteristics is certainly a must. The geographer Turri: "The perception of a country takes place through a series of constituent elements of the territory that impress with their evidence, beauty, grandeur, singularity, or perhaps because they are repeated, as characteristic and unmistakable leitmotifs. These visual elements, detectable in the landscape (rivers, villas, squares, castles, shrines,…), an integral part of the history and culture of the inhabitants, can be called by the term *iconemi*" (Turri 1998).

The value and therefore the enhancement of each cultural knowledge, architectural period and of all creative expression is a must. Recognition also of the "little" things is now necessary. Only in this way it will be possible to transmit a lively perception

Fig. 5 Strada Vittoria, 10 km road built during drainage reclamation interventions, beginning of XX century, Emilia-Romagna. Photo by Chiara Visentin

to the communities that the semantic quality of these *artifacts* and such landscapes is a heritage of civil society (Fig. 5).

Works and places that perhaps have not artistic or natural excellence in their singularity but become fundamental because they are part of a unique and irreplaceable system. Such as the farms of the Agro Pontino or the organization of rural Emilia Romagna and Veneto agricultural districts, deriving from the twentieth-century hydraulic reclamation (Figs. 6 and 12).

Often to affirm the value of a place or an object, it has happened that it was necessary to rewrite a new history, to remove it from paths previously consolidated in memory, an example, among the many, is the current enhancement of the landscape of water drainage, or of military archeology, from the bunkers of the World wars to the lines of defense that collect them. Until a few decades ago, they were considered only a cumbersome unsightly problem on the territories. As for Italian landscape of the twentieth century, no new history must be rewritten. Because it is so evident in its fruitful result, in its landmarks that signal and direct the area, in its daily operations, in its environmental beauty, that it is rather to wonder why even today a national recognition of this heritage is not equated in Italy with that for example in Netherlands.

The identity and physical, immaterial and material characteristics of this territorial systems (most of the time of the past two centuries), arrived almost intact in the contemporary, between architecture, construction of the territory, hydraulic technique, engineering of the machines, are so high that they manage to make us go beyond the actual names and protagonists that they created it, to instead capture the overall size of the asset.

They are real points of reference, very recognizable in an "invisible" horizontal landscape, which makes them emerge in terms of volume and formal characteristics (Figs. 7, 8 and 9).

Fig. 6 Podere 66, Lazio; Ruins of XX century rural farms in East side of Veneto. Photos from Latina Archivio di Stato, fondo Opera nazionale combattenti and by Andrea Morucchio, Chiara Visentin ©

Fig. 7 San Siro water pipe building, San Benedetto Po Mantova, Lombardy. Photo by Giuliano Ferrari ©

Fig. 8 Torre di Fine water pumps building, Caorle Venezia, Veneto. Photo by Chiara Visentin

Expressions of art and culture, but also of the memory of an era, which dictated a good standardization and persistence of architecture and an a-temporal strategic awareness of high territorial planning, to the interventions on the early twentieth-century Italian landscape. Craftsmanship, local and regional tradition, mixed with highly intelligent universal techniques. The original matrix of traditionalism and regionalism issues of Italian modernism (Pagano and Guarniero 1936), which in some cases are very evident in the hydraulic buildings built for the Emilian water reclamation, are for example interesting architectural compositions, in some ways refined in their intrinsically rooted, according to a precise dictate of fascist doctrine, to the value of modernity seen as a historical continuation of tradition. Two different and complementary binaries of ideological communication: an outfitted innovation accompanied by widespread and popular local historical identities (Visentin 2011).

The knowledge of places is therefore fundamental also for the redemption of these territories: lands that we normally consider non-strategic, both for tourism and for environmental development: landscape to be considered *common good* more for an eloquent expression of the identity of a place and its social value, main connection between past and future, then for its visible dimension. For that reason as well as talking about *common good*, we should more often talk for these areas about *common well being* (*common well being* in 2016 was included in Italian law n.163 as a parameter in economic planning: here a special committee has selected 12 indicators of fair and sustainable welfare).

Fig. 9 Mondine water pumps building, Moglia Mantova, Lombardy. Photo by Chiara Visentin

Who writes this essay loves reading literature about places (as about identity and memory), but one of the risks is often the sliding toward a sort of metaphysics of the place taken up in a kind of immobility and historicity (the *soul of places*, the *feeling of places*, *genius loci*, etc.). In order not to fall into this risk, we have to live *into* the places, we must tell about real links that are established with them (Visentin 2008). Landscapes are social constructions, and result of continuous production by the inhabitants. Studying statistical data about Italy we discover that it's a Nation of territorial inequalities, but if we initially suppose that these are well defined between North and South, then we understand that we are wrong: there are gaps both in the North and in the South, both on the mountains and on the flat lands, both along the sea coasts and internal areas, both in rural and periurban areas. A geography where creation of new opportunities not only improves the economic aspect but also the hope, will, and empathy of the communities located in specific places (Fig. 10).

3 Slowness in Ordinary Flat Lands

In his being *ordinary* landscape and not clearly recognizable, the landscape of water drainage reclamation, typical of Northern Italy agricultural flat lands, is instead an extraordinary context: the result of an intense, continuous, and rapid transformation

Fig. 10 Artificial canal, Caorle Venezia. Photo by Andrea Morucchio ©

made mainly from the early twentieth century (Fig. 11). Etymologically we refer to the adjective *extraordinary* in its Latin meaning: *extraordinarius*, extra "out" and ordo-dĭnis "order" (Fig. 12).

According to Treccani Italian encyclopedia "… not ordinary, which goes out of the ordinary, from the usual, from the normal or from the common" (Treccani 2008):

Fig. 11 Roncocesi. Photo by Luigi Ghirri: this is the last photo of the great Italian photographer Luigi Ghirri before he died: taken in Roncocesi, the place he lived near Reggio Emilia, there is only the fog that dissolves the margins and an artificial canal that runs toward infinity

Fig. 12 Women Rice pickers's work break; Construction of drainage canals by "scarriolanti", the workers of drainage reclamation, beginning XX century. Images from Fondo Consorzio agro mantovani reggiano CMR Fototeca Baratta Mantova ©

normal and *common* we could consider the so-called *ante-reclamation* territory, not yet redeemed; in reality it is *communis opinio* that drainage reclamation landscape is instead the truly ordinary landscape, for its total historical and morphological integration in the Italian territorial contexts.

A planning, implementation, and remarkable episode. The reading of this historical and environmental heritage is necessary to identify, through careful investigation, how every so-called "natural" facts and the multitude of "artificial" *artifacts* increase in value as a whole, in their *making system*, in mutual integrated understanding. In a continuous evaluation of the territory that is caught in the slowness of the rhythm of its waters.

Consider this landscape as a cultural heritage, between architecture, history and the environment, thanks to an important heritage and historical memory for the community, associating it in all respects with how it is reflected on "the landscape" by the Italian Constitution (linked in the protection to the "historical and artistic heritage of the Nation"—art. 9), it is to establish which are the rules and the indications for its protection and enhancement.

Many artifacts (buildings, artificial canals and basins, wet areas, rural land quarters, and land infrastructures) of the Italian drainage reclamation were built in the twentieth century: this circumstance tends to be a justification for their actual value, on a declared minor importance compared to older constructions.

Of those elements (whether they are manufactured, hydraulic machines, bridges, artificial canals, expansion basins, etc., that the community sees or goes beyond daily, contexts that are essentially involuntary co-protagonists of their daily life habits) the majority of people don't know their identity and history, nor their purpose and use. A striking example among all: some hydraulic drainage buildings with specific and highly recognizable historical, architectural, and formal characteristics (think of Mondine water pumps in Moglia or San Siro in San Benedetto Po—Mantua, to mention the more manifest), next to which the community lives and works, have never been seen up close for many decades, even just for simple curiosity, until a massive communication and promotion information project of the last 15 years, by the institution that manages them, a Reclamation Consortium of Emilia-Romagna/Lombardy regions.

Imposing landmarks in the landscape that certainly attract but are left there, far from everyday experiences, as if they were large and inactive cathedrals in the desert which in this case is identifiable with the vast horizontal landscape of the plain. Well these "cathedrals" are architectural and hydraulic machines that have been operating for many decades now, today equipped with sophisticated latest generation control mechanisms, necessary to ensure the hydraulic safeguarding of the territory, contingencies for this to be known and understood. The same applies to those highly connotative elements of the agricultural and hydraulic landscape which are the artificial canals and expansion basins: not only controlled extensions to depress the flood levels, but real important naturalistic areas. All human works, necessary, which will continue over time to guarantee the life of the cities and the life of the countryside, allowing irrigation and agronomic optimization of the soil, but which in this new millennium can also be something more: places, territories, buildings to be shared together with the communities adjacent to them, to allow everyone to participate in the enhancement of an articulated territory, between human history, landscape, and nature, as is this portion of Po Valley.

Now, after this information campaign and realized projects of landscape discovery, these common landscapes are knew but fortunately most of the great way directions reserved for fast circulation are little connected with this rural territory, do not affect central areas of these places, continuing to look at them "from afar". If we take as an example the territories of the regions of Veneto (North) and Lazio (in central Italy), the roads leading to the coastal strip of seaside tourism, North-South for the Veneto and East-West for Lazio, are mainly local roads, with contained dimensional characteristics that do not attack the settlement frame composed of agricultural areas, canals, reclamation roads, peculiar to the territories crossed, ensuring that the fast tourist flow can have slower rhythms here. In this landscapes, roads that were once built for the hydraulic redemption and transformed for swamps' cultivation are still protagonists: the structure is clear and strongly identifying, the characterization of road-canal-embankment-bridge-row system *frangivento* (a succession of planting of hygrophilous vegetation, eucalyptus trees for the Agro Pontino, poplars, mulberries, and hornbeams in Veneto) is mostly legible, becoming a patrimonial element of these landscapes.

More than land infrastructures, waterways are an important opportunity in these territories, organized both on the natural network of rivers and on the artificial mesh of the canals. The navigable water routes, certainly to be increased, have already been the theme of recent realizations. Existing river circuits are very interesting, in Veneto on Livenza and on Lemene rivers, but also on drainage reclamation canals, as well as the possibilities that may come from *Pianura Blu*, a work-in-progress project about the waterways of the Pontine district in Lazio, started in 2012, which tends to enhance the active navigable network (on the Ufente, Cavata and Amaseno). Realized in 2013–2015, *GiraLivenza* is a project to promote tourism in the provinces of Pordenone, Treviso, and Venice (a walking, cycling, canoeing, and kayaking, on horseback), the *GiraLemene*, from one year before, is a tourist-historical-naturalistic itinerary for pedestrians and cyclists that, starting from Stalis Mills and arriving at the Port of Falconera, following the course of Lemene river (Visentin and Vanore 2015).

The strong presence of watercourses, lagoon and valley areas in Eastern Veneto, in the Po Valley from Reggio Emilia to Mantua, and lakes behind the Lazio area, plays a role of strategic importance from a socio-economic, cultural, and naturalistic-environmental point of view. Unified ambits of landscape units that preserve an intrinsically cultural connotation, constants identifiable in most of these landscapes that frame them as real patrimonies to be lived in "slow" alternating times and ways of which nature is habitual:

- rigorous linearity in the territorial development of the districts;
- geometric organization of arboreal scenes and boundary lines between rural plots (in close connection with the ancient *centuriatione* system of the ager);
- analogy between the characteristic artifacts of the reclamation and correspondence in the conception of their hydraulic organization on the territory;
- repeated spatial relationship on a large scale of rural farms and farms with the agricultural production context surrounding them (mostly in large estates).

Rhythm, distances, geometry, seriality, volumes, horizontality, and *verticality* of the elements, *measures, style*: these are some of the keycodes through which the shape of this landscape can be deciphered. But above all is *continuity*: an organizational and structural continuity (from water pumps to rural buildings) that dot this new landscape and make it work (Figs. 13, 14 and 15).

The relationship with water is the main element and detector to interpret the identity of these landscapes to be read as expression of a community, result of the spatial modifications made by man over time, by regenerating and culturally reinterpreting the meanings (immaterial) and the signs (materials) that identify it (Fig. 16).

Our entire Italian landscape is historic and modern at the same time and even more so landscape organized by the hydraulic redemption operated in the first half of the last century, which is still operational today (Fig. 17). There are three peculiar elements in the ordinary nature of drainage reclamation landscape. Being *historical*, often very ancient in its formation; also being *modern*, sometimes *contemporary*, in its construction, and finally *operative* in its daily activity, in the productivity of its soil. By participating in the development and enhancement of a territory, not only will it be possible to know it, but above all it will be possible to monitor it, in order to allow its significant use, comprehensibility, and conservation, by reading its functional and cultural varieties.

Fig. 13 Drainage rural lands near Ca' Vendramin di Taglio di Po Rovigo, Veneto. Photo from Google Earth

Fig. 14 Drainage rural lands near Mazzocchio Latina, Agro Pontino Lazio. Photo from Google Earth

The geographical and morphological experience "on the field", understood as observation and identification of the elements have therefore fundamental significance: pertinent to the cognitive investigation, interpretation, and personal description on consolidated bases is necessary thanks to the variety of landscape investigation.

Sharing the territory, planned by active regulations and awareness is fundamental: in fact, it is not enough to preserve the external appearance of the places, restore the historic architectures or retrain the canals; participation in the action takes concrete and factual value as a function of the meaning and uses that are attributed to the

Fig. 15 Drainage rural lands near Saiarino Ferrara, Emilia-Romagna. Photo from Google Earth

Fig. 16 Torrione artificial canal, Gualtieri Reggio Emilia, Emilia-Romagna. Photo by Giuliano Ferrari ©

contexts, in a positive *welfare* that tends, through knowledge, to the active construction of a friendly environment, a sort of relational setting, a *living landscape*, where history and memory can help a correct and operative reading of places and a truthful recording of their future, so that, keeping the memory of the already been, we can evaluate with foresight the negative and positive examples of the past (Visentin and Bonini 2014).

Fig. 17 Graphic redesign by Chiara Visentin of Lucio Gambi, *Drainage Reclamations*, plate 62, in Thematic Atlas of Italy, Italian Touring Club, 1992. Drainage/hydraulic reclamation operations: in *green* from the eleventh to the fifteenth century; in *blue* from the sixteenth to the eighteenth century; in *red* from the nineteenth onwards

The symbolic and real value of the importance of these places, taking up the happy definition of Andréas Rodrìguez-Pose "the revenge of the places that don't matter" (Rodrìguez-Pose 2017), could really be to listen and learn the time-life of these landscapes: the slowness.

4 Conclusion

What has been told here has had important examples among the artificial landscape of the hydraulic reclamations organized throughout the Italian Peninsula (and islands), mainly from the end of nineteenth century to the mid-twentieth. Examples told because it is very present and fundamental in these places the relationship between landscape and community, environment and man, production and life both for urban (and human) and rural settlements (Fig. 18). We can call it a "contemporary" landscape. But that of water drainage landscape is only one of the examples of the multitude of "ordinary landscapes" that dot our beautiful country. The landscapes of the reclamation are perhaps the least *forgotten* ones: their operational function for the community leads them to know and to live in daily more than others that are really in true condition of oblivion. But perhaps starting from the case of artificial water territories of the "production" and management (Visentin 2012), we can learn to see the other landscapes as well, perhaps precisely those that for example cannot even be working, but are forgotten in their times of life and peculiarity that doesn't relate to the routine of the times of man today.

Places and landscapes to be re-evaluated through local cooperation policies: landscape, man/community, action: together they manage to produce value, to produce new perspectives for forgotten areas (Gambino 1997). Together to regenerate and know the existence of forgotten places that are near us, that can be found *around the corner*.

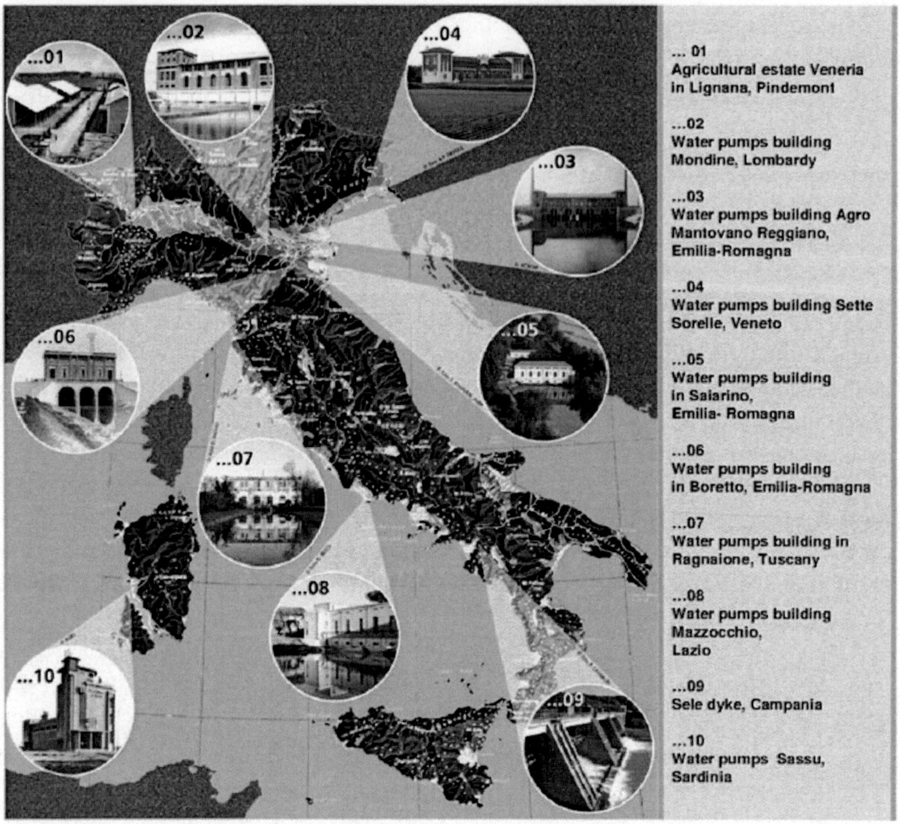

Fig. 18 Italian itinerary through the forgotten places of the hydraulic reclamation. The architectures that identify the landscape. This map summarizes an Academic research on the landscape of Italian hydraulic reclamation (Piedmont, Lombardy, Emilia Romagna, Veneto, Tuscany, Lazio, Campania, Sardinia) edited by Visentin, for Docomomo, a non-profit organization dedicated to documentation and conservation of buildings, sites and neighborhoods of the Modern Movement: *Gente d'acqua. Itinerario attraverso le architetture delle bonifiche agricole in Italia* (*People of water. Itinerary through the agricultural hydraulic drainage architectures in Italy*)

References

A.A. V.V. (2008), Treccani. Vocabolario. Treccani, Roma

AA VV ISPRA (2014) Il Consumo di suolo in Italia, ISPRA Rapporti 195/2014. ISPRA - Istituto Superiore per la Protezione e la Ricerca Ambientale, Roma

Clementi A (ed) (2002) Interpretazioni di paesaggio. Studi metodologici per l'applicazione della Convenzione Europea del Paesaggio. Meltemi, Roma

Donadieu P (2002) La société paysagiste, Actes Sud, Arles (Bouches-du-Rhône)

Galanti G M (1789) Della descrizione geografica e politica delle Sicilie. Gabinetto Letterario, Napoli

Gambino R (1997) Conservare-innovare. Paesaggio, ambiente, territorio. Utet, Torino

Hutabarat Lo R (2009) Walkability: what is it? J Urban: Int Res Placemaking Urban Sustain 2(2):145–166

Magnaghi A (2010) Il progetto locale: verso la coscienza di luogo. Bollati Boringhieri, Milano

Michell D (2003) Cultural landscapes: just landscape or landscape of justice? Prog Hum Geogr 27(6):790

Pagano G, Guarniero D (1936) Architettura rurale italiana. Quaderni Triennale, Milano

Rodrìguez-Pose A (2017) The revenge of the places that don't matter (and what to do about it). Camb J Reg Econ Soc 11(1):189–209

Saramago J (2010) Una terra chiamata Alentejo. Feltrinelli, Milano

Solnit R (2001) Wanderlust. A history of walking. Penguin, New York

Starobinski J (1999) Paysages orientés, in Renzo Zorzi, Il paesaggio, dalla percezione alla descrizione. Marsilio editore, Venezia

Turri E (1998) Il paesaggio come teatro. Marsilio, Venezia

Turri E (2003) Il Paesaggio degli uomini. La natura, la cultura, la storia. Zanichelli, Bologna

Visentin C (2008) L'architettura dei luoghi. Principi ed esempi per una identità del progetto. Il Poligrafo, Padova

Visentin C (2011) Il Patrimonio architettonico e ambientale dei paesaggi della Bonifica. Valorizzazione e promozione della memoria dei luoghi. Consorzio di Bonifica dell'Emilia Centrale Editore, Reggio Emilia 2011

Visentin C (ed) (2012) Gente d'acqua. Itinerario attraverso le architetture per le bonifiche agricole in Italia. Supplement to Docomomo Italia J 31, XVI year, Docomomo, Roma

Visentin C, Bonini G (2014) Paesaggi in trasformazione. Compositori, Bologna

Visentin C, Vanore M (2015) Heritage of Water. Patrimonio e paesaggi di Bonifica. Istituto Alcide Cervi Editore, Gattatico (Reggio Emilia)

Food on the Bike

Massimo Montanari

Abstract Food is a key asset for the italian tourism but speed is not the best tool to discover it. Because of high-speed's tale of the modern society, people are discouraged from learning the fascinating relationships between the biodiversity of local food and landscape. Walking and cycling can help people pay more attention on local food as a legacy of a peculiar landscape.

Widespread throughout the country, Italian culture—including gastronomy—has its distinctive character in diversity and variety. The umbilical link with the territory (always expressed in the plural: individual territories) makes Italian culture unique, and travel indispensable to meet products, recipes—but before that, landscapes. In fact food, before being on the table, manifests itself in the landscape, and if food is local, the landscape is what we go through. The landscape of wheat, wine, oil; the landscape of meat and milk, fruit and vegetables, without forgetting the landscapes of woods and water, and even those of the food industry, which designs new landscapes and launches new challenges to the environment. Landscapes—not always beautiful—that for the most part are men's constructions, "artificial" products that interact with "nature". Giacomo Leopardi explained it very well: "A very large part of what we call natural is not; indeed it is rather artificial: as to say, worked fields, trees and other plants educated and arranged in order, rivers narrowed within established boundaries and addressed to a given course, and similar things, do not have the appearance they would naturally have. So that the sight of every country inhabited by any generation of civilized men is an artificial thing, very different from what it would be in nature" (In praise of Birds).

Looking at these landscapes and understanding them, seeing where nature ends and artifice begins, means to reflect about our history. It means to understand that beauty arises when nature and artifice go hand in hand, geography and history harmonize; and that, in the course of history, landscapes have had (and often still

M. Montanari (✉)
Università di Bologna, Bologna, Italy
e-mail: massimo.montanari@unibo.it

P. Pileri and R. Moscarelli (eds.), *Cycling & Walking for Regional Development*, Research for Development, https://doi.org/10.1007/978-3-030-44003-9_7

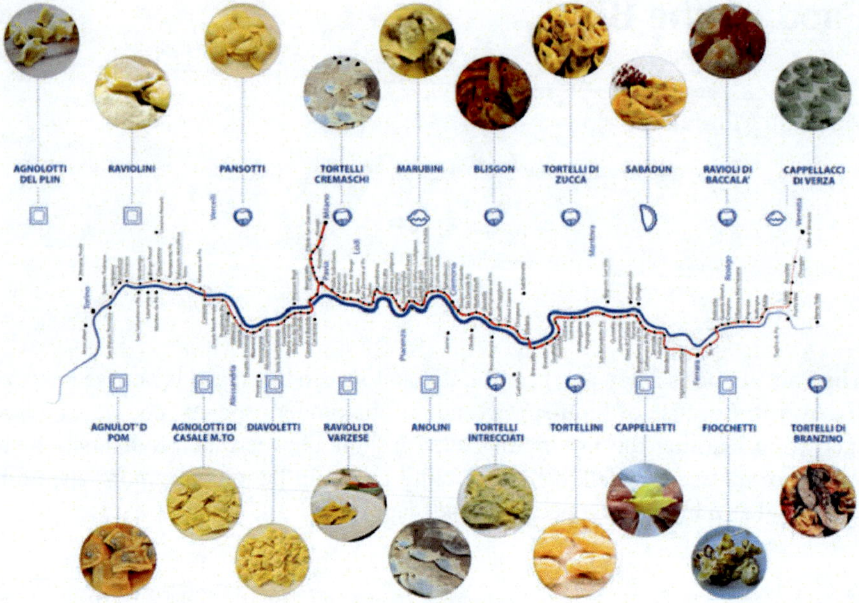

Fig. 1 A slow line can mend together what we dont'see as a sequence of local stories along time, anymore. Here the case of a peculiar Italian pasta (tortellini family) that along the Po river had an important story that a cycle tourism track like VENTO can emphasize and show to slow travellers *Source* VENTO Reserch Group 2018

have) a fundamental reason why they exist: producing food for survival, and for everyday pleasure.

Understanding landscapes is the real challenge facing today's tourist, became a high speed and hurriedly traveller who crosses spaces he does not know and does not realize what they mean for him, anymore. If something is broken in the deep bond between man and food, this can perhaps be reconstructed starting from the places we go through every day and that can talk to us, tell stories, explain things that still have space in our memory. It's only a matter of reactivating it, starting from the observation that food culture is too often and too uniquely oriented on the plate, on the shop, on the restaurant rather than on the landscape. Travelling is necessary, and looking at landscapes, and understanding them, and above all understanding that those landscapes are not something different from us, but actual creations and, literally, men's inventions. Understanding them—just like a work of art, a monument, a square we pass through—will be the first step to, perhaps, conceiving new ideas on how to preserve or improve them, or, at any rate, live them consciously.

Crossing landscapes, then, but how? By car, speeding through the guard rails of a motorway? On foot, with that slow pace that perfectly allows us to enjoy each detail, but more painstakingly to grasp the whole? An ideal vehicle exists: the bicycle, which follows our pace and anticipates it, allows us to run and stop, giving us the freedom of a conscious and intelligent movement. Let's rediscover it (Fig. 1).

The Role of Historic Roads to Preserve and Valorize the Landscape

Alberta Cazzani and Maurizio Boriani

Abstract Historic roads are roads that, through design, experience or association, have contributed to our culture in a meaningful way. Historic roads form a very interesting architectonic, technical and cultural linear system: not only the traces, but also the road features (walls, bridges, tunnels, drain wells), the connected buildings (churches, chapels, fortifications, custom-houses, mills, forges, furnaces, mines) rose out of ancient religious, military, commercial or industrial functions with a relationship between villages, towns, landscapes. The type of road, its history and current condition determine the most appropriate action for preservation. As with all historic resources, it is impossible to separate a historic road from its context, its setting. Leaving out the more ancient trails (roman and medieval roads) with a lot of archeological importance, there are many roads dating from the eighteenth to the early twentieth century now transformed or abandoned. These can now be used as landscape, and cultural resources, but it is necessary to preserve and conserve this heritage with specific inventories, analysis and restoration policies and practices in order to define management and valorization plans with historic, touristic and ecological purposes.

1 Why Historic Roads Are Part of Our Cultural Landscape

The study of the problems of preservation and conservation of the historical, architectural and landscape heritage has led, after a long debate, to a more extensive concept of the terms monument and landscape. Developing the definition of monument from a building of considerable historical-artistic interest, to historic center, rural architecture, industrial archeology and the idea of landscape from a panoramic view of "significant beauty" to "an area as perceived by people, whose character is the

A. Cazzani (✉) · M. Boriani
Politecnico di Milano, Milan, Italy
e-mail: alberta.cazzani@polimi.it

M. Boriani
e-mail: maurizio.boriani@polimi.it

P. Pileri and R. Moscarelli (eds.), *Cycling & Walking for Regional Development*, Research for Development, https://doi.org/10.1007/978-3-030-44003-9_8

result of the action and interaction of natural and/or human factors" (The European Landscape Convention 2000).

At the same time, historical research has added another element to the study of the most important events, the "material culture", that means signs and actions documenting the productive, social and religious aspects of a civilization.

Historical and natural sites are considered as resources, assigning them not only a cultural interest, but also an economic value and social utility.

Because of a new perspective of the value of the past, the preservation domain is being enlarged by a radical increase in the types of artifacts being considered worthy of protection and conservation. In addition to monumental high-style architecture—traditionally the preservation concern—whole new categories of sites are now recognized as equally meritorious: cultural and natural landscapes, historic towns and villages, vernacular and industrial structures, XX Century architecture. The current preservation approaches are focused on analyzing materials and building techniques, conserving the historic stratification of the architectural and landscape heritage and defining compatible reuses.

Preservation is no longer the study of an individual monument, but a practice that deals with the historical heritage in its entirety, involving the set of relationships that each component establishes with other components and the context.

The heritage inherited from the past must be considered as a system of parts strictly and inseparably connected to each other and the landscape must be approached as a built, man-made territory, which man has lived in and modified over the centuries (Roe and Taylor 2014).

The historical road system is a fundamental element of this landscape (Boriani and Cazzani 1993). The ancient roads are in fact closely linked with the geographical and geomorphological territorial system and are part of a collective project of the landscape's transformation. New settlements, development of production, military or political purposes and commercial trade always required communication and connections (ICOMOS 2008).

The historic road network was much more widespread than today: even areas now marginalized by economic development (as mountain pastures, abandoned quarries and mines, woods and low productivity agricultural areas) had to be made accessible at that time due to their economic importance. Other areas considered strategic (like mountain passes, ports, ancient state borders), today have lost much of their importance but were in the past served by a dense road network.

The construction features and technique of the historical road system were very different from the current ones: transport took place on foot, on horseback or by animal pulling carriages. Travel was heavily influenced by weather and by political, economic, fiscal and safety conditions. Often the road consisted of a "bundle of roads", distant from each other, which were alternately used according to the convenience of the moment.

Road sections were rarely more than 4/5 m wide and as narrow as 2 m. The roads were much more related to the natural conformation of the land and the hydrographic conditions and much less to the need to shorten travel times; the construction materials used were always local.

The historical road system has often left significant traces: the route layout, the features and the strong punctual elements (like bridges and other engineering works, customs, hospices, religious buildings). This is a system that without continuous use it is no longer maintained and develops decay.

It is essential to consider cultural heritage both the road itself (its layout, its construction techniques), and the system of engineering and architecture works that support it (like retaining walls, bridges, fords, pavement, drainage system, memorial stones, fountains, customs, inns, chapels, trees).

However, it should be noted that, compared to other architecture of the past, historical roads suffer from technological obsolescence due to the industrial revolution. An ancient building, if well preserved, still offers many possibilities of use, while an ancient road, built for now obsolete transport systems, today offers little economic interest, in relation to significant maintenance costs.

The ancient roads are subject to a rapid process of decay with obsolete segments abandoned and transformed by enlargements, rectifications, new pavement. The abandoned sections, left without maintenance, decay and this heritage quickly disappears.

Where the historic road is still in use, some features and connected buildings (as post stations, inns, chapels, customs) can be preserved or converted to other uses, but the bridges and the tunnels are unable to withstand intensive use by modern form of transportation and require replacement.

However, the historic roads that are now obsolete due to modern traffic needs are an important resource for slow mobility networks, supporting tourism, cultural and environmental projects.

Furthermore, often the residual parts of the historical road system across mountains are more subject to the action of natural agents: their decay is therefore accelerated by the encroaching vegetation, by rainwater, by geological phenomena (like landslides and avalanches).

Only the main buildings, the most important road engineering works and the most resistant to natural phenomena roads maintain the testimony of past constructive knowledge. Other traces are also relevant, but the connection of them in a continuous network has often been lost and so has the possibility of recognizing their meaning and importance as a system.

It is therefore worthwhile to recover ancient roads and their contexts not only because they are part of our historical heritage, but because making them still passable allows us to discover the ways in which we related to that territory in the past, both from an economic and cultural approach.

2 Conservation and Rehabilitation of Historic Roads: Criteria and Goals

Over time humans have traveled from one place to other for many ends, for example, to make a living, to exploit natural resources, to carry or bring messages or to visit certain places. We know that from the beginning, societies have tended to create land routes for the movement of goods and individuals. This trail network represented the communication and the movement of many different people and languages through the centuries.

Along the ancient roads, it is still possible to find areas and elements of landscape significance, historical and cultural interest, that have been marginalized by new roads and new urban centers. For this reason, important resources have not yet been compromised by industrial development and intense urbanization.

Therefore it is important to prepare conservation and rehabilitation plans aimed at re-establishing a link between the architectural and environmental assets of the historical landscape, to ensure an overall reading of the territorial stratification and particularly its conscious and congruent use.

With the exception of the ancient Roman routes and of some famous pilgrimage trails, historical roads are still often not considered as part of cultural heritage.

The historical road system and the sites connected to it are instead important resources that deserve to be valued for various reasons.

First it is necessary to avoid the loss of a precious document of our past, significant for cultural, technological, social, economic and political history. In this sense, the ancient roads can now be interpreted as true open-air linear museums: along their traces it is in fact possible to recognize the complex signs of man's presence in the area: the places of living, the exploitation of natural resources, agricultural and manufacturing production, faith, trade, politics, war, but also places of natural history and geography.

Furthermore, a modern country needs to control its entire territory extensively: widespread ancient roads represent an already available and well-distributed system of accessibility even to the most marginal areas. This ancient road network can support the problems of forest management, water regulation, fire control and, more generally, to the ecosystem governance.

Finally there are the opportunities offered to environmental and cultural tourism, which can enhance the economic development of these marginalized areas that can present their value today as refuges from the intense modernization of our cities and countryside (Hernández Ramírez 2011).

It is therefore important to develop programs aimed at valorizing the ancient roads as crucial elements in the construction of the territory, landscape and economic development. The historic road should not be considered only as a cultural asset to be listed, but as a site of cultural production to be lived and enhanced. Historic roads give cycle-pedestrian excursions and alternative tourism opportunities to discover natural habitats and areas which are rich in historical evidence.

In some cases—such as for the complex road system of transhumance in southern Italy along the Apennines Tratturi (further detailed in part 3.1 of this paper)—conservation and rehabilitation programs have created linear parks, in which the road is the element that helps visitors connect with historical, cultural sites, architectural and natural-landscape (Latini 2000; Carnevale 2005; Sarno 2014).

It is necessary to consider historic roads in local and regional planning tools: for example in Italy, the Emilia Romagna Region since the 1990s has developed detailed studies on historic roads as part of provincial and regional planning (Devoti 2007).

The Landscape and Urban Plans should not only indicate the historical road network, but also define strategies for their conservation, their recovery as cultural assets (as part of our cultural heritage) and their use as cultural routes, hiking and cycle-pedestrian trails.

Historic roads represent an important resource for the development of cultural and ecotourism.

The adequate preservation of historic roads involves more than an appreciation for history. Historic roads constitute one of the most difficult resources to preserve. By their nature, they generally traverse great distances and include a broader contextual landscape. The ability to preserve, maintain and protect the integrity of the corridor through which they travel is important. They represent a resource that, in many instances, is still functioning as originally designed (Fig. 1).

Specifically in the United States, the National Park Service (NPS, the Federal agency in charge of landscape and cultural heritage preservation and management) considers and plans the National Scenic or Historic Trails and Historic Roads not only for their track, but also for the areas, sites, views, natural aspects connected to them (CRM 1997; Lay 2006; Marriot 2010; McCahon et al. 2012).

Since the 1990s, the NPS has introduced the concept of Historic Transport Corridor (CRM 1993) that "is a historically significant route along which people and/or goods have moved, in which there is evidence that the natural environment has been modified by mankind. It is a linear cultural landscape, that combines cultural values with the natural ones. Some corridors may have ceased to evolve, although evidence of their existence may still be discernible. Others may continue to function up to the present time, and indeed the evidence of their historical evolution may well have survived, although possibly in modified form through continuing use" (CRM 1993: Cameron, p. 5).

One of the goals of the concept of the historic corridor is the conservation of the historical resources, both natural and cultural that exist along its length; these are aspects that give it a special character. The composite is more than the isolated parts, as it includes the historic sites, ancient structures, archeological sites, old tradition, people knowledge and important ecological zones. A historic corridor includes individual components that enrich it profoundly.

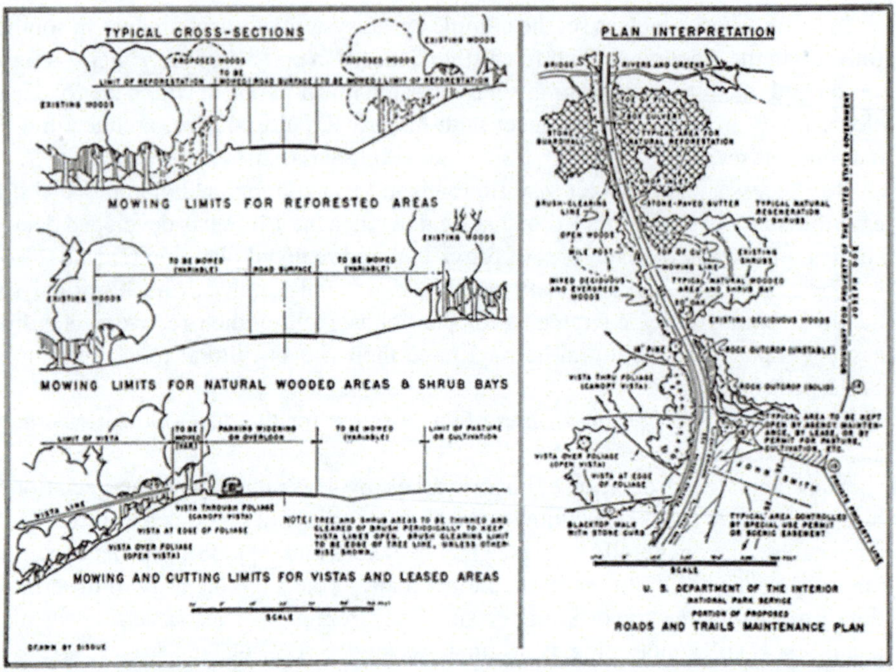

Fig. 1 U.S. National Park Service, Roads and Trails Maintenance Plan: constructive and perceptive landscape analysis of an historic parkway and its context in order to define conservation and valorization treatments (McCahon et al. 2012)

This means that a historical road is protected, recovered and enhanced at the same time with a buffer zone, sometimes extended for several miles in width, going to constitute a truly linear park that connects other areas and sites of natural or historical interest.

The analysis and study of the historic roads concerns not only the road track, but also the connected areas, valuating their landscape, historical, natural and cultural significance and the educational-recreational potential (Klein 2015) (Fig. 2).

However, correct planning of the context of historic roads is not sufficient. The old roads suffer from their intrinsic delicacy, which requires special care and attention: the pavement, the features, the tree and green furnishings, the engineering works, are essential components of the road: their disappearance or replacement with modern materials produces an irreparable loss of historical-documentary values and of environmental quality.

In this regard, considering the system of historic roads, Switzerland has demonstrated in the last 30 years that to establish intervention criteria and to define preservation and management recommendations for landscape planning at different levels, it is always necessary to carry out primarily a knowledge phase with a large-scale inventory and a subsequent analysis and detailed survey, in order to identify components, materials and construction techniques, assess the state of decay and/or problems of

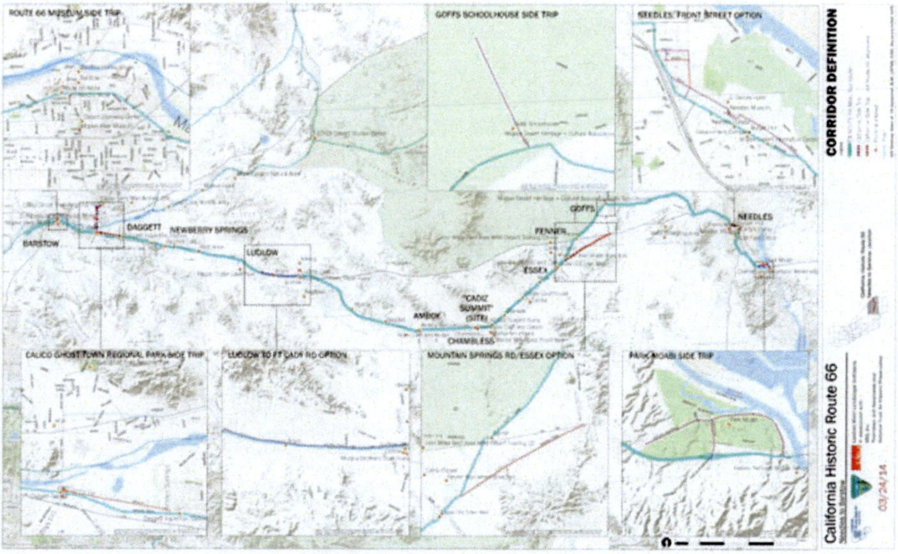

Fig. 2 California Historic Route 66: corridor definition to better preserve the historic roads and the linear landscape connected with it (Klein 2015)

use and conservation to delineate the necessary recovery and enhancement measures (Colombo and Bellini 2003; Aerni et al. 2003) (Fig. 3).

It should also be remembered that one of the fundamental conditions for preservation of an ancient road is linked to the maintenance of continuity travel: the frequent passage guarantees a periodic monitoring of the necessary maintenance, often performed by the passers-by themselves. A frequent trampling, for example, is an important aid to contain the growing vegetation on the road surface.

In order to establish a safeguard policy, it is essential to take into account not only the characteristics, importance and significance of a historical road, but also the level of current conservation, evaluating the historical substance or permanence, considered as the value of the residual road construction and the surrounding landscape. This concept has been very well defined and considered by preservation policies of historic roads in Switzerland since the 1990s (USTRA 2010).

The level of residual historical substance of the various tracks identified, allows us to evaluate how much of the ancient artifacts is still conserved today, to be able to establish priority levels for preservation measures.

The evaluation of the historical significance of the roads, the level of conservation of their construction features and the quality of the landscape they pass through are the condition to define a recovery and enhancement plan. In this way, a mule trail of local importance from a commercial-economic point of view can acquire an interest in safeguarding as evidence of the permanence of particular artifacts (as interesting structural engineering works; characteristic surviving features or decorative elements) and vice versa a road of national historical importance that has undergone intense alterations and transformations (rectification of the route, construction of new

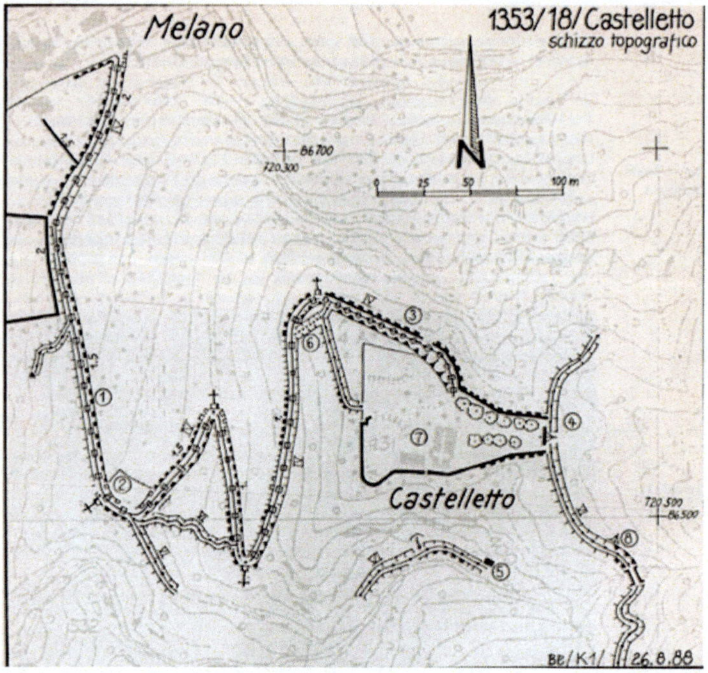

Fig. 3 Preservation policies of historic roads in Switzerland: a detailed survey to analyze architectural components, materials and construction techniques (Colombo and Bellini 2003)

infrastructures, enlargements, deviations, new modern materials) that have deleted the historical value, will have a low level of conservation and therefore a lesser need to be protected, at least in its material aspects.

It is necessary to evaluate the integrity of the historic road and to evaluate its context before making recommendations for its preservation. If the road and the context are in place, but sufficiently degraded, they must be rehabilitated more than conserved.

The preservation project must therefore define the list of necessary interventions: the planning of the territorial context of the road, the conservation and recovery of the most significant artifacts along the road or in the immediate vicinity, the conservation and maintenance of the road and sites/features related and also the program of future management and maintenance criteria with the indication of the compatible functions, as well as the definition of facilities and use promotion.

In this regard, it is necessary to analyze the uses and potentials of the historical road, the economic and social benefits and to evaluate the impact of the future services and infrastructures, in order to conserve the historical and natural values.

It is important to preserve the cultural heritage of the corridor and to encourage the study of the historic road, as well as to document its effect on the natural resources and culture of the era, and to promote the preservation of the natural and historical resources that survive.

A preservation and reuse plan for a historic road is able to promote the economic growth at the local level, as well as to stimulate the community development, and at the same time to promote national and international cultural tourism that is sensitive to conservation.

The ancient roads can therefore be read both as a support to the new ecological and natural management needs of the landscape, and as a real linear museum where natural sciences, rural and vernacular tradition, culture and great human history and educational policies are inextricably intertwined.

3 Preservation and Reuse of Historic Roads in Italy: From Historical Investigation to Valorization for Eco and Cultural Tourism

3.1 Research and Projects on Historic Roads in Italy in Recent Years

Italy is a largely mountainous country: the Alps in the North, separate Italy from France, Switzerland and Austria; the Apennines divide the Po Valley from the Tyrrhenian coast, making the connections between the Tyrrhenian Sea and the Adriatic Sea difficult. For this reason in Italy, a dense network of mountain passes have developed since ancient times: over 2000 m in the Alps and over 1500 m in the Apennines.

Even excluding Roman-medieval roads, whose parts are still often visible, many traces from the eighteenth–nineteenth century still exist. They have been adapted only partially to motorized viability. The roads built in pre-industrial times conformed to the original topography and orography; for the more recent roads instead it was preferred to cross valleys by building bridges and viaducts and to tunnel through mountains, abandoning the old and difficult paths.

In Italy in the 1980s–1990s, there was a new focus on historic roads, with a growing awareness of the historical value, trekking potential and environmental valorization. These activities fostered the conservation and reuse of the ancient roads and trails.

Over several years, studies were conducted and included—between many others—the Simplon (Lodi 2005; Caneparo et al. 2006; Tognozzi 2017), Spluga (Planta 1993; Wanner 2006; Jäger 2016; Fay 2018), Stelvio (Jochberger et al. 2000; Silvestri 2001), Gries (Zanzi and Rizzi 1998), Septimer (Monteforte and Ceretti 1995), San Marco (Via Priula) (Cellini and Guglielmi 1987) Passes in the Alps; Abetone and Tambura (Via Vandelli) (Gambi et al. 1987; Pellegrini and Pozzi 1989) Passes in the Apennines.

Other studies explored coastal roads of the main lakes of northern Italy, in particular the Regina road on the western coast of Lake Como (Frigerio et al. 1995), the road built by the Austrians along the Eastern coast of the same lake (Brembilla and Denti 2018) and the Gardesana, a parkway along the West coast of Garda Lake (Mazza 1984; Cazzani 2003; Cavallini 2005; Zezza 2009).

Studies for less important, but no less significant itineraries concerned the Salt Roads connecting Liguria with the Po Valley and other coastal or mountain trails, as well as mule tracks serving pastures.

Finally, on the 100 year anniversary of the First World War, new research was dedicated to the military road system built to access the Alpine front and to the defense features connected to it (Sigurtà 2017).

These works also defined interventions to rehabilitate disused and abandoned roads and to restore road facilities (bridges, tunnels, retaining walls, water networks) and road support buildings (like churches and chapels, customs, inns and taverns).

Often, as many ancient roads are now included in National or Regional Parks, restoration projects have been linked also with large-scale plans for the conservation and management of the connected areas of historical or natural interest to promote cultural and ecotourism.

These studies and plans were usually followed by the publication of excursion guides and detailed cartographies, as well as by inventories of the historical, landscape and natural assets connected with the different trails. It would be too long to describe even a selection of these works here: for more details, please refer to the bibliography.

3.2 Reusing Historical Roads as Slow Mobility Networks to Discover the Italian Landscape: Some Significant Initiatives

Considering the numerous initiatives carried out in recent decades in Italy, it seems important here to describe some examples in which the ancient roads have been studied and restored analyzing the historical value of the route and related features, buildings and areas, but also the significance of the ancient trails in the current policies of landscape planning and promotion of slow mobility.

Since the 1990s, interest in cycle and pedestrian excursions has increased considerably not only in mountain areas, but also in lowland areas and in periurban zones. Walking and cycling are not only good for our health, but also allow us to visit places of natural and historical interest, to discover ancient traditions and to appreciate food and local wine varieties.

In this regard, the most famous European walking route crossing Italy is the Via Francigena—a system of paths used since ancient times by pilgrims—that connects England with Rome (Stopani 1988, 2010).

The Francigena is an itinerary that has changed over the centuries, due to natural conditions, different territorial boundaries, tolls, safety reasons and presence of bandits.

The reconstruction of the "real" track of the Via Francigena would be today an impossible challenge, because it never actually existed: it is reasonable however to connect the main buildings and places related to the wayfarers along the Via.

A new focus on the Via Francigena developed in recent times as a result of the rediscovery during the 1970s and 1980s, of the Saint James roads system. For the 2000 Jubilee, studies, publications and research were dedicated valorizing this important European pilgrimage route, including the program defined by the Italian Ministry of Tourism in 1993 in collaboration with some Italian Regions and Municipalities. Approved by the European Commission as part of the EU Action Plan, this program led to the recognition of the Via Francigena as a "Cultural Route of the Council of Europe" in 1994.

The Via Francigena is not a restored historic road, but it is a complex itinerary that connects important cities, villages and monuments and crosses scenic areas, sometimes using ancient traces, promoting slow mobility. The original religious purpose of the Via Francigena is now shared with trekking and cultural tourism intentions (Baruffi 1999; Alberti 2005; Centini 2011; Del Boca and Moia 2015; Ardito 2010; Ardito 2016).

Here it is more important to describe some projects where the conservation and reuse intervention has specifically addressed historic roads, considered for their typological and constructive characteristics.

It is noteworthy to underline the historic roads inventory that the Politecnico of Milan made with Lombardy Region Forestry Agency in 1990s, comparing nineteenth-century military maps with current maps to understand the permanence of the itineraries. This inventory connected with the Swiss one, because many Lombardy roads were built to link Italy with Switzerland and it was essential to have a common investigation tool (Cazzani and Sangiorgio 2015).

A detailed analysis and survey work have been made in some Lombardy areas to check the conservation levels of the inventoried roads and the architectural and landscape features connected with them, considering decay and alteration problems, architectural and natural values and rehabilitation potentials. A GIS about Lombardy historic roads was set up to organize and join all the gathered and surveyed data and to manage them (Fig. 4).

These studies developed guidelines for preserving and rehabilitating the Lombardy historic roads system and a project to valorize and reuse the First World War road system as a trekking trail and an open-air museum.

The historic road system of Lombardy Region is now included in the Rete Escursionistica della Lombardia—REL, (Lombardy Trekking Network, issued by the Lombardy Region Law, n.5/2017) that is focused on the promotion of cultural, natural and landscape heritage. The main goals of the REL are to develop the attractiveness of rural areas, to enhance hiking and mountain activities, increase eco-compatible tourism and support maintenance interventions of trekking paths, with particular regard to the historic trails and roads.

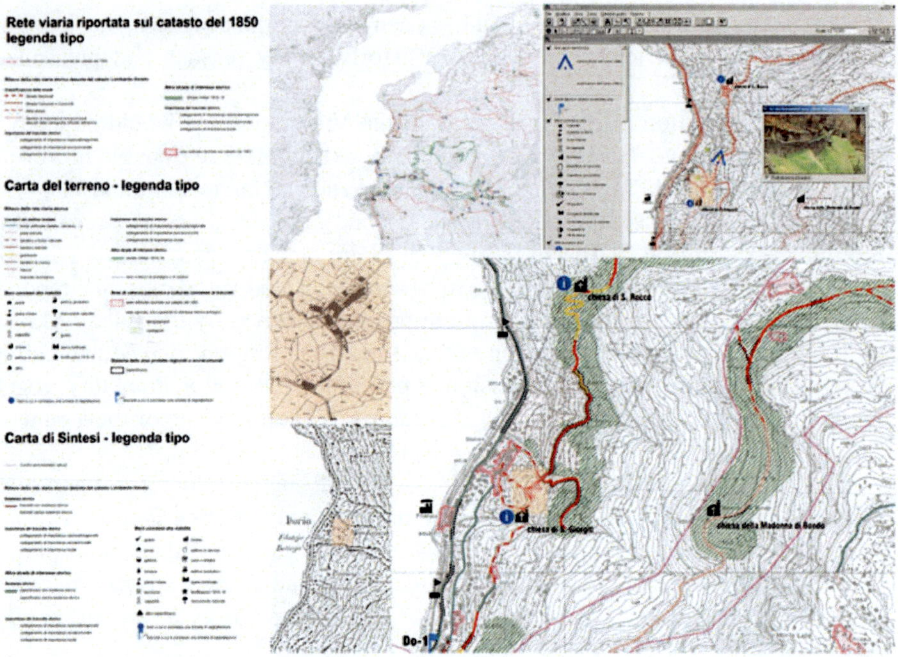

Fig. 4 An example of Lombardy Region historic roads survey, studying old maps, considering the conservation level and analyzing historic roads materials, characteristics and features. All this information was organized to set the related GIS (Cazzani and Sangiorgio 2015)

The First World War road and defensive system has been preserved and restored also in Piedmont and in Trentino Adige. From 2010 to 2013 the *Forti e Linea Cadorna. Dai sentieri di Guerra alle strade di pace* plan was developed. It focused on surveying (setting a web GIS tool), conserving and valorizing the monumental complex of trenches, roads and fortifications along the Cadorna defensive line on the border between Italy, Austrian-Hungarian Empire and Switzerland. Historic roads, military components and features have been surveyed and restored in order to make the past comprehensible to future hikers. The final goal was the definition of the *Slow Panorama project*: a slow mobility trail—connected with the Val Grande National Park—creating a bike park, involving historic and landscape resources (Trotti 2013; Vaschetto 2015; Viviani and Corbella 2017).

Another important project preserves and valorizes the First World War roads and fortifications system, the *Peace Trail* (*Sentiero della Pace*), an alpine hiking trace 500 km long, connecting the Stelvio Pass with the Marmolada Mount. Since the 1990s the Trentino preservation office has inventoried First Word War heritage in order to conserve and restore military roads, trenches, walkways, cornerstones, forts and connected features to delineate a cultural and trekking trail. The *Sentiero della Pace* gives a complex experience: a magnificent alpine scenario—partially included in Stelvio National Park—and a tangible and intangible historic system well explained and promoted (Fabbro 2016).

Considering the Apennines roads it is interesting to remember the Via Vandelli, an ancient commercial and military road, recently preserved and reused as a linear historic and natural park.

The Via Vandelli was built between 1738 and 1752 to connect the city of Modena (today in Emilia Romagna) with Massa (today in Tuscany), giving an outlet to the sea of strategic importance for the Modena Ducat. The designer was the abbot, topographer and mathematician Domenico Vandelli, who made a direct connection through the difficult Tambura Pass at 1670 m.

When the Duke of Modena lost control over the city of Massa, other routes were built, allowing an easier crossing of the Apennines and this road lost much of its interest, both military and commercial. For this reason, it was never modernized and adapted to the needs of motorized traffic.

Today it remains a magnificent example of an ancient road, well preserved, especially on the Tuscan side, where there is an impressive series of steep hairpin bends supported by monumental dry stone walls (Fig. 5).

The road is included in the Tuscany Regional Park of the Apuan Alps that defined management program in collaboration with the Municipality of Massa and received funding from the European Union. They have recently launched a series of conservation and valorization interventions to reuse the historic road and its features, as a significant component of the Regional Park landscape and historic heritage. The works were completed in 2006 and today the road is a true open-air museum

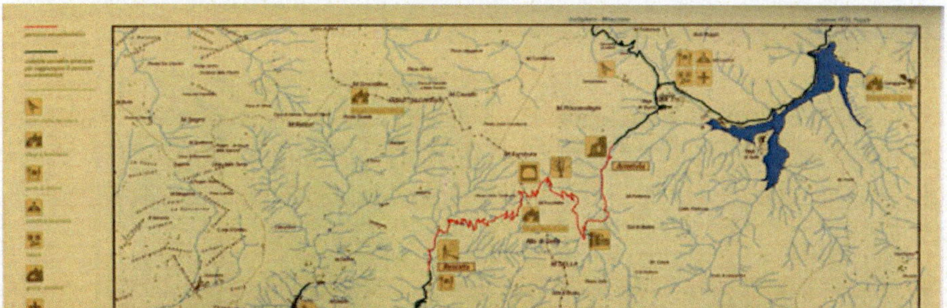

Fig. 5 Via Vandelli—The road from Vagli to Massa, in red the restored hiking-open museum trail (Giovannetti and Puccini 2006)

Fig. 6 Via Vandelli linear open museum: information panel explaining the construction features of the historic road (Giovannetti and Puccini 2006)

displaying construction techniques, as well as areas of archeological and environmental interest linked with it. This walking tour is now an amazing experience: hikers can explore a beautiful landscape and clearly understand the cultural significance of a well preserved historic road (Giovannetti and Puccini 2006; Ferrari 2019) (Fig. 6).

As mentioned above, one of the most significant examples of preservation and reuse of historic roads in Italy concerns the transhumance network. It is important to remember that transhumance is a very ancient form of pastoralism or nomadism organized around the migration of livestock between mountain pastures in warm seasons and lower altitudes the rest of the year. Most peoples who practice transhumance also engage in some form of crop cultivation, and there is usually some kind of permanent settlement. Particularly in South of Italy (in Abruzzo, Molise and Puglia Regions) several studies and plans have been set up and the Tratturi network is now considered as a cultural, environmental and scenic integrated resource and an opportunity for ecotourism and sustainable development (Aloi et al. 2007; Pellicano 2012; De Santis and Cutino 2013; Cucciolla and Muratore 2014; Russo and Bourdin 2016) (Fig. 7).

Grassy paths—up to 111 m wide—were connected with furnishings and fittings that shepherds used during their journey, like stone edges, bridges, dry stone walls,

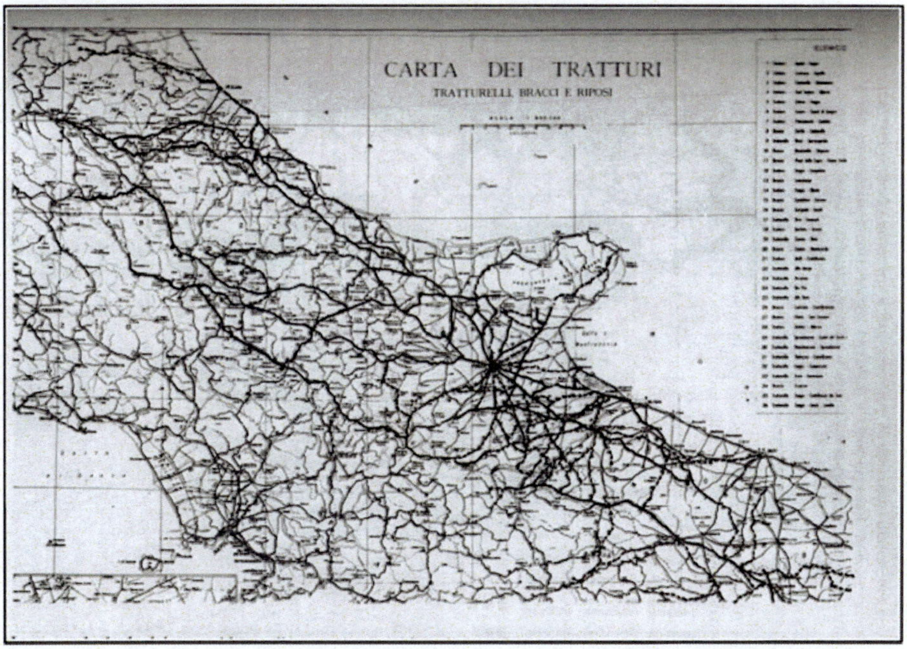

Fig. 7 The Tratturi network: grassy paths mainly placed in the Regions of Abruzzo, Molise, Puglia, Basilicata and Campania (map by Commissariato per la reintegra dei tratturi di Foggia 1959)

fountains, troughs, stables, taverns, churches, chapels, trade and commercial buildings. The Tratturi, particular in Molise, Abruzzo and Puglia Region are still partially conserved, covering 1300 km and linking different protected areas, included the Gargano, Abruzzo and Majella National Parks.

These green trail ways are State properties: they are still conserved because they have been included in a multi-level preservation system. Since 1976 the Tratturi have been listed by the National Preservation Law n.1089/1939 as part of the archeological, historical and cultural Italian Heritage; in 1997 a Regional Law was issued in Molise to protect the Tratturi network and the Regional Park of the Tratturi was established (Figs. 8 and 9).

The Tratturi symbolize a relationship to the past and to local identity, an archeological and historical tangible and intangible heritage to be conserved, safeguarded and valorized.

Several studies and plans—often carried out in cooperation with different Regions and supported by European Union funding—made the restoration of the Tratturi possible defining rural greenways, bike trails, horseback riding paths, open-air linear museums, in order to valorize a complex system of natural and cultural landscape, historic heritage, traditional agriculture and food.

Walking along the Tratturi allows individuals to discover important monuments and archeological sites, enjoying scenic landscapes, learn about ancient tradition and taste gastronomic specialties.

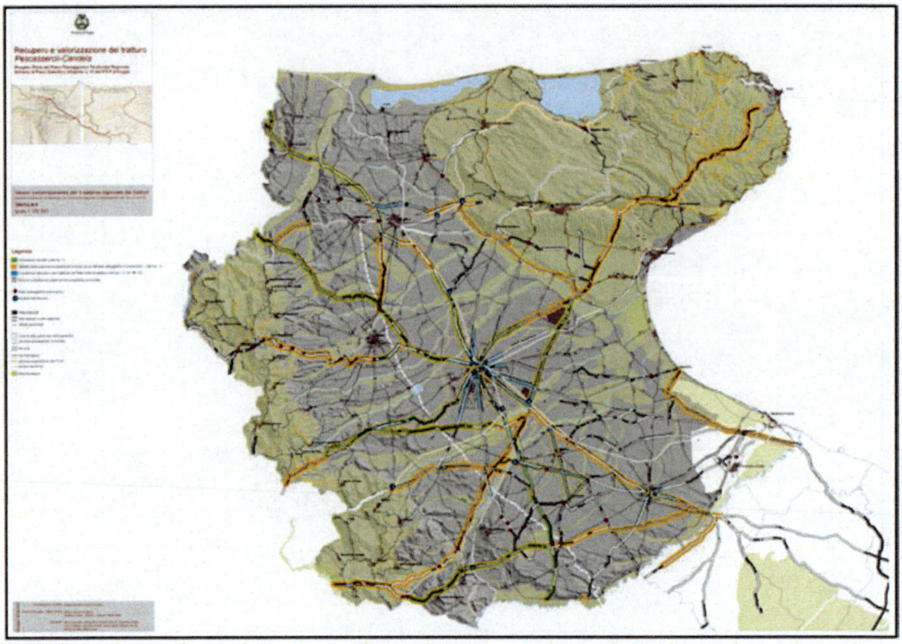

Fig. 8 Preservation and valorization of the Tratturo Pescasseroli-Candela (Foggia), management strategies. The Tratturo is considered a greenway linked to Regional and National Parks and related to a slow mobility network (Cucciolla and Muratore 2014)

The Tratturi represent a new approach of landscape conservation and tourism promotion of the local territories and communities. The Tratturi are not only physical traces and tangible resources, they are also an intangible heritage characterized by knowledge, practices and interaction models. For this reason in 2018, the Transhumance practice, as a network of European pastoral areas, was submitted to be part of the Unesco World Heritage List of Intangible Heritage.

Trans regional and international plans and programs issued thematic trails, museums, visitor centers, involving local people, promoting traditional products, setting educational courses and increasing job opportunities. The Tratturi network is very well integrated with planning tools and preservation policies, demonstrating how historic roads can represent a strong settlement for active landscape conservation, valorization and management.

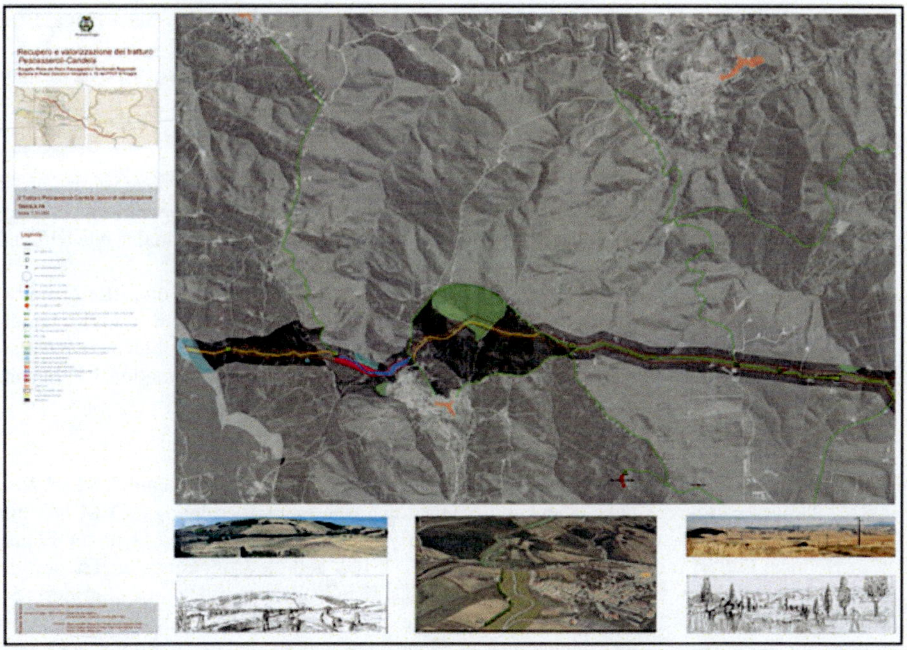

Fig. 9 Preservation and valorization of the Tratturo Pescasseroli-Candela (Foggia), enhancement actions. The Tratturo is an eco-corridor connected with agricultural areas, woodlands, historic sites, ancient features. The goal is to define a linear cultural and scenic landscape to promote trekking, tourism and educational programs (Cucciolla and Muratore 2014)

Author Contributions The present paper was conceived and written jointly by the two authors. In particular, Maurizio Boriani developed part 1 and 3.1, Alberta Cazzani developed part 2 and 3.2. The authors collaborated in the writing of the abstract and for the final revision of the manuscript.

References

Aerni K, Bitz V, Benedetti S (2003) Historische Verkenhrswege im Kanton Wallis. ASTRA OFROU Berna

Alberti A (2005) I Sentieri lungo la Francigena. Da Siena a Roma. RAI ERI, Roma

Aloi E, De Castro MG, Zollo A, Guarino N (2007) La rete tratturale come mosaico paesistico ambientale ed opportunità di ecoturismo. In: Agribusiness Paesaggio & Ambiente, vol X, No 3

Ardito F (ed) (2010) La Via Francigena. Touring Club Italiano - TCI, Milano

Ardito F (2016) Via Francigena. Touring Club Italiano - TCI, Milano

Baruffi GA (1999) La Via Francigena. Sulle vie dei pellegrini in provincia di Pavia. Provincia di Pavia

Boriani M, Cazzani A (1993) Le strade storiche. Un patrimonio da salvare, Guerini, Milano.

Brembilla G, Denti M (2018) La strada stretta tra i monti e il lago. Vicende e suggestioni della Strada Militare Lacuale da Lecco a Colico (ex Statale 36). Editrice Associazione Culturale Luigi Svanagatta, Varenna

Caneparo B, Ceria E, Destefanis A (2006) La via del Sempione da Domodossola a Briga. Eventi & Progetti, Biella
Carnevale S (2005) L'architettura della transumanza. Indagini, tecniche costruttive, restauro. Palladino, Campobasso
Cavallini G (2005) La strada nella roccia. Uomini e vicende nella storia della viabilità del Garda Occidentale. Fondazione Negri, Brescia
Cazzani A (2003) La Gardesana Occidentale: una parkway italiana da conservare e valorizzare. In: Guerci G, Pelissetti L, Scazzosi L (eds) Oltre il giardino... Le architetture vegetali e il paesaggio. La storia e la percezione del paesaggio. Metodologia, strumenti, fonti per l'analisi e la lettura del paesaggio. Olschki, Firenze, pp 129-144
Cazzani A, Sangiorgio C (2015) Inventory, preservation and valorization of historic roads in Lombardy Region (Italy). Current policies and future plans. In: Toniolo L, Boriani M, Guidi G (eds) Built heritage: monitoring, conservation, management. Fondazione Politecnico di Milano, Springer Cham, Heidelberg, New York, Dordrecht, London. Springer International Publishing, Switzerland, pp 69–79
Cellini P, Guglielmi F (1987) La Strada Priula. Editrice Cesare Ferrari, Clusone
Centini M (2011) La via Francigena. Xenia, Pavia
Colombo M, Bellini G (2003) Le vie di comunicazione storiche nel Canton Ticino. USTRA Berna
CRM - Cultural Resource Management (1993) Historic trasportation corridors. In: CRM—Cultural Resource Management, vol 16, No 11. Particularly: Cliver EB, A New and Dynamic Element of Heritage Preservation: 1, 8; Morton T.B US/ICOMOS and the Conference on Trasportation Corridors as Cultural Landscapes: p 4; Cameron C, The Challenges of Historic Corridors: 5–8; Liebs CH, Reconnecting People with Place. The Potential of Heritage Transportation Corridors: pp 9–11; Nowak T.R. Techniques of Identifying and Evaluating Corridors and Trails: 12–13; Krakow JL. Identifying and Evaluating Historic Corridors and Trails: 14; Leach SA, HABS/HAER Documents Automotive Corridors 26–27
CRM - Cultural Resource Management (1997) The national trails system. CRM cultural resource management NPS national park service, vol 20, No 1
Cucciolla A, Muratore G (2014) Recupero e valorizzazione del tratturo Pescaseroli-Candela. Progetto Pilota del Piano Paesaggistico Territoriale Regionale. Schema di Piano Operativo Integrato n.10 del PTCP di Foggia. Relazione generale. Provincia di Foggia, Foggia
De Santis P, Cutino N (2013) Tratturi e transumanza. Profili fra passato, presente e progettualità future. Wip edizioni, Bari
Del Boca L, Moia A (2015) Sulla Via Francigena. Storia e geografia di un cammino millenario, Utet, Torino
Devoti S (2007) Piano Territoriale di Coordinamento Provinciale. La Viabilità storica. Provincia di Piacenza
Fabbro C (2016) La grande guerra e il sentiero della Pace. Reverdito, Trento
Fay MC (ed) (2018) Donegani, l'ingegnere tra le Alpi. Lo Spluga, un passo verso l'Europa. Società Economica Valtellinese, Sondrio
Ferrari G (2019) La Via Vandelli. Antica strada, nuovo cammino. Edizioni Artestampa, Modena
Frigerio G, Luraschi G, Martello Frigerio D (eds) (1995) L'Antica Via Regina. Tra gli itinerari stradali e le Vie d'acqua del Comasco. Società Archeologica Comense, Como
Gambi L, Minghelli F, Pellegrini M (eds) (1987) La via Vandelli. Strada ducale del '700 da Modena a Massa – I percorsi del versante emiliano. Artioli Editore, Modena
Giovannetti L, Puccini R (eds) (2006) Sulle Alpi Apuane nel Settecento. La Via Vandelli e il Casone di Ripanaia: storia, archeologia e restauro. All'Insegna del Giglio, Firenze
Hernández Ramírez J (2011) Los caminos del patrimonio. Rutas turísticas e itinerarios culturales. In: PASOS Revista de Turismo y Patrimonio Cultural, vol 9, No 2, pp 225–236
ICOMOS - International Scientific Committee on Cultural Routes (CIIC) of ICOMOS (2008) The ICOMOS charter on cultural routes. ICOMOS
Jäger G (ed) (2016) Der Splügenpass. Zur langen Geschichte einer kurzen Transitraoute – Il Passo dello Spluga. La lunga storia di una breve via di transito. Verlag Bündner Monatsblatt, Chur

Jochberger W, Gerd P, Tappeiner KG (2000) Die stilfser Joch Strasse-La strada del passo dello Stelvio-the Stilfser Joch road. Associazione turistica gruppo Ortles parco nazionale, Athesia, Bolzano

Klein J (ed) (2015) California Historic Route 66. Needles to Barstow Corridor Management Plan. Bureau of Land Management California Desert District California Historic Route 66 Association

Latini M (ed) (2000) Abruzzo. Le Vie della Transumanza. Carsa Edizioni, Pescara

Lay M (2006) Preserving historic roads. In: Routes roads, vol 334, pp 88–91

Lodi V (2005) Sempione. Strada napoleonica, galleria ferroviaria, ferrovie e tranvie nel Verbano-Cusio-Ossola. Alberti, Intra

Marriott PD (2010) The preservation office. Guide to Historic Roads, The James Marston Fitch Charitable Foundation New York

Mazza A (1984) Il Meandro: una strada per il Garda occidentale, Grafo, Brescia

McCahon ME, Sutherland L, Shaup S (2012) Design and management of historic roads. NCHRP-National Cooperative Highway Research Program Web only Document 189

Monteforte F, Ceretti L (1995) Septimer. Il Pass da Sett. Lyassis Edizioni, Sondrio

Pellicano A (2012) Geografia e storia dei tratturi del Mezzogiorno. Ipotesi di recupero funzionale di una risorsa antica. Aracne, Roma

Pellegrini M, Pozzi FM (ed) (1989) La via Vandelli. Strada ducale del '700 da Modena a Massa – Dal Frignano alla Garfagnana e al Ducato di Massa, Artioli Editore, Modena

Planta T (1993) Le antiche strade dello Spluga. Edizioni Centro Studi Valchiavennaschi, Chiavenna

Roe M, Taylor K (2014) New cultural landscapes. Routledge, London, New York

Russo S, Bourdin S (2016) I tratturi fra tutela e valorizzazione, Grenzi, Foggia

Sarno E (2014) La cartografia storica tratturale per lo studio dei paesaggi della transumanza. Un caso di studio. Bollettino AIC, No 150

Sigurtà D (2017) Montagne di guerra, strade di pace. La Prima Guerra Mondiale dal Garda all'Adamello: tecnologie e infrastrutturazioni belliche. Franco Angeli, Milano

Silvestri I (2001) Le strade dell'Umbrail e dello Stelvio dal Medioevo al 1900. Tipografia Pradella, Bormio

Stopani R (1988) La via Francigena. Una strada europea nell'Italia del Medioevo. Le Lettere, Firenze

Stopani R (2010) L'altra Francigena. La quotidianità del pellegrinaggio medievale. Le Lettere, Firenze

The European Landscape Convention (2000) Florence Italy (https://rm.coe.int/1680080621)

Tognozzi C (2017) Le vie del Sempione. Storia, arte, immagini, documenti, itinerari, racconti, cartografia. Macchione Editore, Varese

Trotti A (2013) Alla scoperta della Frontiera Nord. Otto spunti di turismo storico-militare tra Varese, Como, Lecco, Sondrio e Canton Ticino. Progetto For.Ti-Linea Cadorna, Museo della Guerra Bianca in Adamello, Temù (BS)

USTRA, CFMS, CFNP (ed) (2010) La conservazione delle vie di comunicazione storiche. Guida tecnica d'applicazione. Aiuto all'esecuzione per il Traffico lento, No 8, ASTRA Berna

Vaschetto D (2015) Strade e sentieri della linea Cadorna. Itinerari storico-escursionistici dalla Valle d'Aosta alle Alpi Orobiche. Edizioni Capricorno, Torino

Viviani A, Corbella R (2017) Tutta la linea Cadorna. Storia Architettura Armamenti. Itinerari di un museo all'aperto. Macchione editore, Varese

Wanner K (2006) Lo Spluga. Il passo sublime. Edizioni Centro Studi Valchiavennaschi, Chiavenna

Zanzi L, Rizzi E (1998) Gries, da Milano a Berna una via per l'Europa. Fondazione Enrico Monti, Anzola d'Ossola

Zezza F (2009) La strada Gardesana occidentale tra la progettazione del primo '900 e la trasformazione di fine secolo. Grafica 5, Arco (TN)

Cultural Heritage Preservation and Territorial Attractiveness: A Spatial Multidimensional Evaluation Approach

Alessandra Oppio and Marta Dell'Ovo

Abstract The introduction of the concept of sustainable development in the field of cultural heritage preservation has stressed the importance of a holistic approach. Achieving a balance among cultural significance retention and economic development is a challenging goal, even more for fragile and vulnerable contexts with limited economic and social resources, low return expectations and a huge tangible and intangible cultural heritage. Given such a complexity decisions about where to place valorisation interventions with the purpose to activate synergies with existing projects and trigger economic and social development processes require to be based on robust evaluation methodologies. According to this instance, Spatial Multicriteria Analysis (SMCA) can support decision-makers along all the steps of the decision process, moving from the intelligence phase to the design and, finally, to the choice one. Within this approach, the study has been focused on the intelligence phase, in order to define a multidimensional analytical framework aimed at mapping widespread cultural heritage with a special attention to its territorial features. The proposed methodological evaluation framework points out the challenge of structuring a decision problem related to the inner areas regeneration by the reuse of cultural heritage placed along slow mobility routes. The results are value maps that provide recommendations for placing culture heritage preservation and reuse interventions, meant as territorial catalyst.

A. Oppio (✉)
Department of Architecture and Urban Studies (DAStU), Politecnico di Milano, Milan, Italy
e-mail: Alessandra.oppio@polimi.it

M. Dell'Ovo
Department of Architecture, Built Environment and Construction Engineering (DABC), Politecnico di Milano, Milan, Italy
e-mail: marta.dellovo@polimi.it

P. Pileri and R. Moscarelli (eds.), *Cycling & Walking for Regional Development*, Research for Development, https://doi.org/10.1007/978-3-030-44003-9_9

1 Cultural Heritage as Territorial Capital

According to the regional development theory, cultural capital can be thought as one of the elements of Territorial Capital. Starting from the first OECD study (OECD 2001) this notion encompasses all those assets which are embedded in the territory and it acts as driver of territorial growth. Within this perspective, cultural heritage can play a crucial role, since by investing in cultural heritage, local bodies can improve the attractiveness of their assets to tourists, and in this way increment the influx of people who spend their own resources (time and money) to visit the area (Giaoutzi and Nijkamp 2006). Cultural heritage has been more and more interpreted as a central matter for urban regeneration and revitalization (Rodríguez-Pose and Vilalta-Bufí 2005; Bowitz and Ibenholt 2009; Leriche and Daviet 2010; Oppio et al. 2015).

By summing up concepts coming from a large number of different but interconnected strands of literature, Roodhouse (2007) has provided a conceptualization of territorial capital elements according to two main dimensions, namely, the materiality (tangibles, intangibles and mixed) and the rivalry (public, private and mixed).

In the classification by Camagni (2009) cultural heritage is mostly a tangible good and an impure public good. In fact, although cultural heritage, in its many aspects, is often freely available or publicly owned, opportunistic behaviours, unsustainable uses and congestion can make its fruition not completely un-rival.

For these reasons, cultural heritage encompasses different aspects, some of which are tangible (such as architectures) and others are intangible (such as the sense and symbolic value attached by local people or tourists, due to history and their personal experiences). At the same time, some specific goods are privately owned and not open to the public and yet their presence can still help inducing as a sense of belonging among local communities. According to the broadly accepted European preservation principles and standards (ICOMOS 2018), recognition of cultural heritage as a common good shall be a precondition of quality interventions.

In line with this recommendation, valorisation interventions are those ones, even though punctual, provide multiple and positive externalities by bringing benefits across economy, culture, society and the environment, as well as by retaining ideals, meanings, memories, traditions and specificities.

Given such a complexity, decisions about where to place valorisation interventions with the purpose to activate synergies with existing projects and trigger economic and social development processes require to be based on robust evaluation methodologies. According to this instance, Spatial Multicriteria Analysis (SMCA) seems to be a promising decision support methodology to help decision-makers along all the steps of the decision process, moving from the problem definition to the final choice. Monetary evaluations fail to capture social and ethical aspects, as well as distributional impacts (Chan et al. 2012). They are based on the aggregation of utility across individuals (Pearce and Nash 1981), a situation which is highly unlikely for heterogeneous groups of people (Munda 2004).

In order to explain the potentials of this approach, the paper introduces the issue of slow tourism and cultural heritage starting from a recent national initiative (§2).

The multi-methodological approach is then described with respect to its operational phases (§3) and in the last section some preliminary conclusions are drawn (§4).

2 Slow Tourism and Cultural Heritage: The "Cammini e Percorsi" Initiative

The program "Cammini e Percorsi",[1] in line with the wider plan "Valore Paese—Dimore" enhanced from 2013 by the Agenzia del Demanio—a government institution focused on the management and enhancement of public real estate assets—aims at the valorisation and reuse of public buildings to host services for walkers, pilgrims and cyclists along interregional and regional routes (Fig. 1).

Given the involvement of Ministry of Infrastructure and Transport (MIT) and Ministry of Heritage, Culture and Tourism (MiBACT), the initiative has been launched with the purpose to strengthen cultural tourism and the network of sites of historical and natural interest by soft mobility and slow tourism.

Landscape and heritage preservation, as well as the reuse of the public buildings located close to historical religious and cycle paths, are considered as the drivers of economic regeneration, especially for inner areas. To achieve these objectives the Agenzia del Demanio has defined many agreements with institutional bodies, such as municipalities and regions, responsible for spatial planning and management.

The first operational step has been addressed towards the selection of routes according to two main categories: historical-religious itineraries and cycling/walking trails.

Nine routes have been recognized as historical-religious itineraries:

- via Francigena is an itinerary of about 1,900 km, which involves more than 244 municipalities, from Canterbury to Rome;
- via Appia is an itinerary of about 600 km, which crosses three regions: Lazio, Campania and Basilicata;
- Cammino di Francesco is an itinerary divided into 21 stations for more than 450 km which crosses three Italian regions: Toscana, Umbria and Lazio;
- Cammino di San Benedetto is an itinerary for more than 400 km across the Umbria, Lazio and Campania regions;
- other minor routes are Cammino Celeste, Cammino Micaelico, Regio Tratturo Magno, Regio Tratturo Pescasseroli—Candela, Via Lauretana.

The cycle-pedestrian routes, on the other hand, are mainly the National Cycleways, i.e., extra-urban infrastructures, often located in areas of historical-cultural or naturalistic value designed for cycling tourism and leisure time. Paths recognized as cycle/pedestrian are as follows:

[1]https://www.agenziademanio.it/opencms/it/progetti/camminipercorsi/.

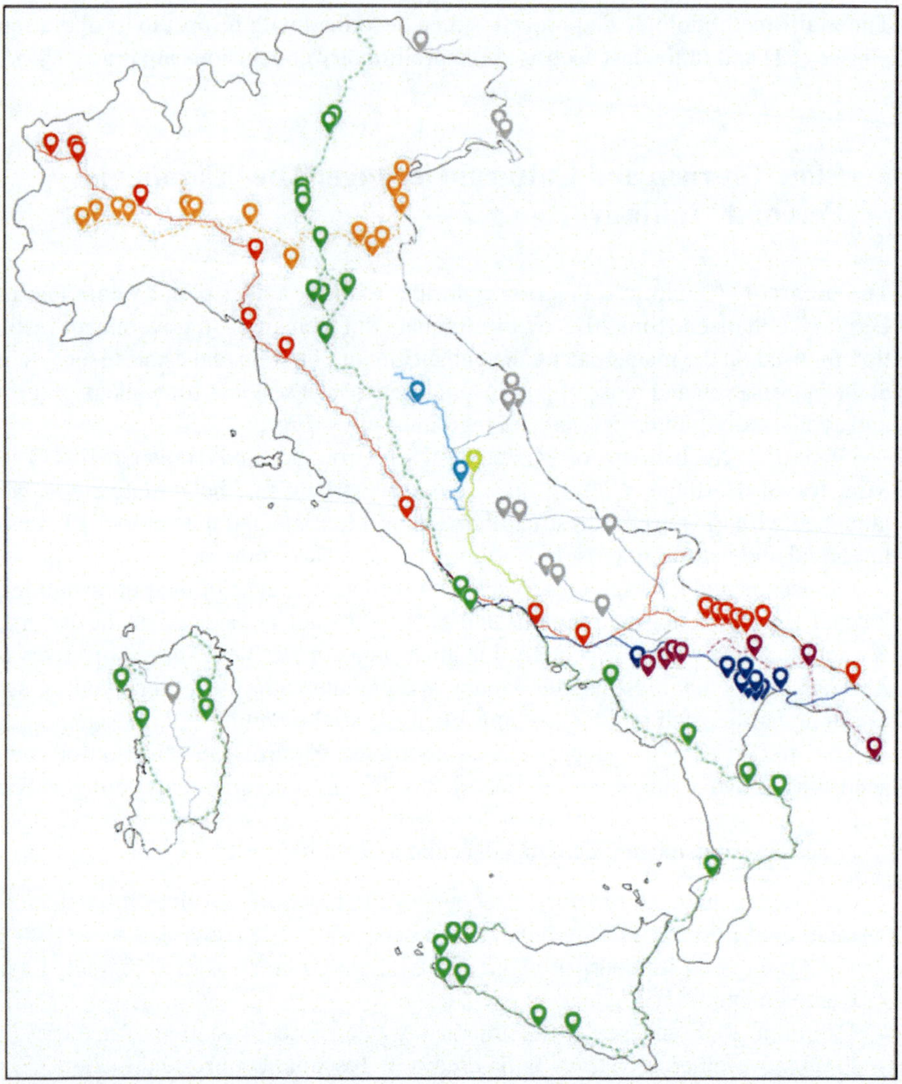

Fig. 1 Routes and properties selected by the program "Cammini e Percorsi". Legend—green: Cicoplista del Sole; orange: Ciclovia Vento; purple: Ciclovia Acqua; red: Via Franchigena; blue: Via Appia; light blue: Cammino di Francesco; yellow: Cammino di San Benedetto; grey: other minor routs. Adapted from https://www.agenziademanio.it/export/sites/demanio/download/agenzia_a_l/Cammini-e-percorsi-cartella-stampa.pdf

- Ciclopista del Sole is an itinerary of about 3.000 km, which crosses 12 regions from Trentino-Alto Adige until Sardinia;
- Ciclovia Vento is an itinerary of about 680 km which connects Venice to Turin by running alongside the Po river;

Fig. 2 Example of public assets 1. Torre Angioina, Atella (PZ); 2. Torre della Bastiglia, Serramazzoni (MO); 3. Torre Angellara, Salerno (SA) (Maps data: Google Maps)

- Ciclovia Acqua is an itinerary of about 500 km, which crosses 3 regions Campania, Basilicata and Puglia;
- other minor: Ciclovia Alpe Adria, Ciclovia Adriatica.

The properties selected to enhance the cultural tourism are disused public assets—owned by the Government or other bodies, such as Regions, Provinces, Municipalities, Anas S.p.A.—and located along the itineraries described above. They belong to two main macro-categories (Fig. 2):

- "traditional heritage": traditional and very frequent assets placed along travel and communication networks. A distinction is made between:
  - typical buildings of the local traditional settlement (e.g., villages, houses, schools, farms, farmhouses, farmhouses, farms, mills, shelters, forest houses, etc.)
  - artefacts linked to the local infrastructure systems (e.g., hydraulic toll booths, road toll booths, cantonal houses, small stations, railway toll booths, etc.)
- "valuable cultural heritage": assets of great cultural interest that stand out in the surrounding landscape for their historical and artistic value, as well as for their size (e.g., castles, forts, fortifications, barracks, villas, etc.).

In 2017, the Agenzia del Demanio has carried on a deep analysis to select the assets to be valued and reused considering both technical and administrative instances as well as urban regulations. At the end of the process, 103 properties have been selected with reference to the distance from the slow routes as a prerequisite: a buffer zone of 5 km for historical-religious itineraries and 15 km for cycle-pedestrian routes. After this phase, a public consultation has been promoted by the Agenzia del Demanio able to involve almost 25.000 participants, of which 20% foreigners and 80% were Italians. They were asked, for example, to elicit their opinion about potential activities to integrate along these itineraries and routs, main important elements able to affect the tourism attractiveness and to identify assets that deserve to be preserved and valorised.

As a result, 43 properties have been submitted to for a first call and 48 for the second. By the end of 2017, 47 proposals have been received for the first call and 13 have been accepted, while for the second call, closed in May 2018, 3 proposals only have passed the selection. Actually, there are no monitoring information about the state of these properties.

With reference to the first call (Free Concession Bid 2017), although focused on widely acceptable principles, some criticalities deserve to be pointed out both in the bids and in the information memorandums provided as guidelines for the proposals. About the bids, the proposals have been assessed with reference to four main criteria (Reuse intervention—max 40 scores; tourism attractiveness—max 30 scores; Environmental sustainability—max 15 scores; Energy efficiency—max 15 scores). The documents arranged by the Agenzia del Demanio had not adequately deepened the definition of the above-mentioned criteria neither how to achieve their underlying objectives. Furthermore, despite the attention to the territorial features of the assets under evaluation, as it emerges from the information memorandums, the bids hadn't considered contexts' strengths and weakness as crucial for addressing reuse interventions.

The methodological approach described in the next section has been defined with the aim of filling the gaps in the current valorisation process, as it has been set up by the Agenzia del Demanio. More in deep, a hybrid methodology based on the combined use of Geographic Information Systems (GIS) and Multicriteria Analysis (MCA) has been suggested in order to support the decision process since its early stages and during its fulfilment. By the use of spatial criteria and utility functions, the multi-methodological approach can be used by the stakeholders involved in the valorisation process—in different ways according to the different role they play—(i) to support the selection of the assets to be reused according to the location features (potentials and weaknesses); (ii) to evaluate how much the punctual intervention on an asset is able to boost synergies with the surroundings sites; (iii) to provide a sound basis for evaluating the impacts of the valorisation interventions across a multidimensional perspective.

3 Multi-methodological Approach

According to the definition of Decision Aiding (Roy 1996, 2005), complex decision problems entail multidimensional evaluation approaches in order to address choices towards a balanced trade-off between conflictual goals. Firstly, the economic regeneration of inner areas by reuse interventions of selected cultural heritage placed along slow mobility routes asks for mapping both the territorial attractiveness and the vulnerability. Secondly, it requires to define local interventions with catalytic effects on the overall territorial development according to economic, social and environmental instances. What is challenging is to find out a balanced reuse strategy, able to bring benefits both to the most attractive sites as well as to the most fragile ones according to the line perspective, that is quite different from a punctual intervention as well from a network of valorisation actions (Moscarelli et al. 2017; Ingold 2015).

Given these premises, the Multicriteria Analysis (MCA) is an evaluation approach able to establish preferences about multiple options according to a multidimensional set of decision criteria.

Differently from Cost Benefit Analysis, MCA doesn't require to assign a monetary value to every input. Within a MCA process all relevant viewpoints are taken into account in order to set up a coherent family of criteria (Roy 1999).

Furthermore, each criterion is selected so that the options' perfomances can be evaluated by their own measurement units that should be clear to all the actors involved and appropriate to the viewpoints considered. Such an evaluation framework is open to discussion about the diverse consequences to be investigated in order to shed light on the decision problems and it could be adapted to instances emerging during the evaluation process (Damart and Roy 2009). Thus, MCA acknowledges value pluralism and supports a reasoned debate around reasons for and against proposals and their underlying beliefs (Saarikoski et al. 2016).

Finally, uncertainty, inaccuracy and arbitrariness, that inextricably affect evaluation, are made explicit by using difference and preference thresholds (Bottero et al. 2019).

Among the many MCA techniques, especially Spatial Multicriteria Analysis (SMCA), by integrating the spatial functions of GIS (Borrough et al. 2015) with the MCA features provides a wide and integrated knowledge about territorial variables (Ferretti 2012; Dell'Ovo et al. 2018; Oppio and Bottero 2018), useful to evaluate the strengths and weaknesses of territories crossed by slow mobility lines.

SMCA is based on the combination of GIS and MCA aimed at exploiting the potentials of GIS in the context of decision-making (Malczewski 2000). SMCA considers both geographic data and decision-makers' preferences as inputs in order to obtain information for decision-making process and an overall assessment of the decision alternatives. The main difference between SMCA and MCA consists, in fact, in the presence of spatial components (Ferretti 2011) useful when the decision problem under investigation has direct effect on the territory. SMCA results strategic in the field of land management, land use planning and more in general spatial planning (Dell'Ovo et al. 2018; Torrieri and Batà 2017) and facilitates the communication of the results through the visualization of maps.

The methodological approach defined to support the decision problem is a multidimensional analytical framework, as represented in Fig. 3. The first phase is related to the analysis of territorial strengths, in order to provide both the overall and the partial attractiveness maps. This value maps are the result of the SMCA.

The second phase of the evaluation aims at developing a territorial analysis to detect weaknesses and to provide an overall vulnerability map.

The third phase combines the results obtained from the previous analysis in order to prioritize interventions on cultural heritage with the support of a priority matrix.

The aforementioned phases will be illustrated in the remaining part of the section.

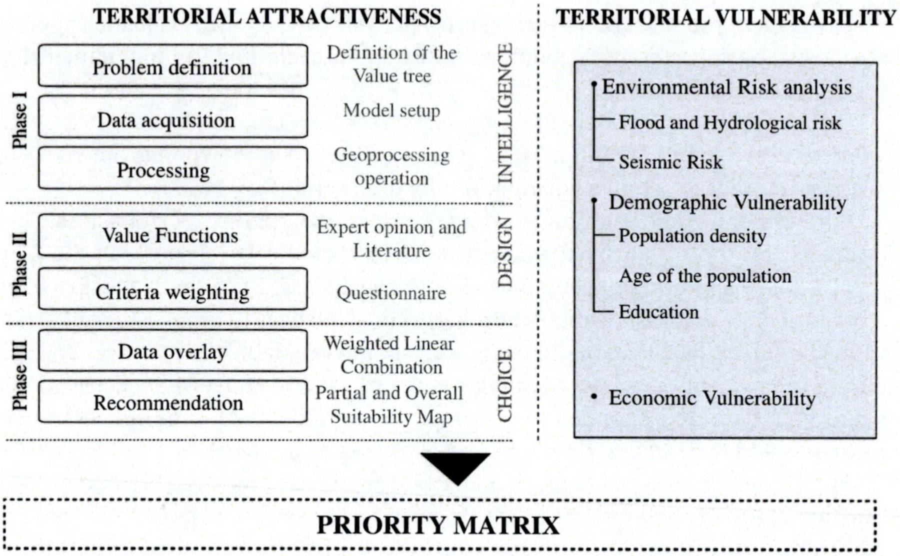

Fig. 3 The spatial multidimensional analytical framework

3.1 The Spatial Multicriteria Analysis: Operational Phases

The SMCA approach can be mainly divided into three subsequent steps, namely, intelligence, design and choice (Simon 1960), that can be described as follows:

- the intelligence phase refers to the description of the decision context and the definition of main objectives. Here the model is framed and data regarding the decision problem are collected and processed;
- the design phase consists of the development of value functions with the support of experts and the criteria weights elicitation;
- the choice phase deals with the aggregation of data previously collected and weights elicited and the interpretation of the obtained results.

3.1.1 Intelligence Phase

The first step consists in structuring the problem and the decision-making framework, by defining the objectives to be achieved in the context under evaluation and the definition of an appropriate set of spatial criteria to be measured. Since this first phase is aimed to define an attractiveness index for territories crossed by cycling and walking itineraries, the problem definition phase has been started from a literature review on this topic, the investigation of similar case studies and the analysis of the outputs of a public consultation carried out in 2017 by the Agenzia del Demanio through questionnaires administrated to 24.632 potential slow trails users.[2]

More in detail, the Scopus database has supported the phase of literature review, and the analysis carried on aimed at understanding the methodology applied and to select a set of criteria consistent with the objective (Hoang et al. 2018; Oppido et al. 2018; Quattrone et al. 2017; Diti et al. 2015; Černá et al. 2014).

For what concerns the case studies, six European itineraries have been examined with the aim of setting up common drivers able to impact on territorial attractiveness.

Considering the public inquiry, it has allowed to explore the interests of the slow paths potential users and to underline their preferences and priorities.

The decision three presented in Table 1 is the result of the above-mentioned analysis.

Once the set of criteria has been framed, it is possible to go on with the available data collection (Source Maps) and to start the geo-processing operational phase. GIS data are strictly related to the selected indicators, as they allow to understand how the sub-criteria are going to be measured. Data can be found on national, regional and local geo-portals by exploring existing databases or they can be developed by detecting and elaborating information provided by official websites or by mapping the territory by the mean of a direct investigation. The geo-processing operations are related to the type of information required, as well as to the selected indicators. As a consequence, they can range from Euclidean Distance; Slope; Density functions; etc. Through these geo-processing operations Source Maps are transformed into Criterion Maps, that show specific information as interval, classes or ranges.

3.1.2 Design Phase

Standardization is a fundamental phase to set different data with respect to a common measurement scale. Once Source Maps have been geo-processed according to the operations identified as the most suitable (Criterion Maps), it is necessary to transform them in comparable units through standardization, that is a mathematical representation of human preferences (Keeney 1992). According to the software used and to the geo-operations applied to the original data (distance; density; etc.), it is possible to assign a dimensionless score (0–1; 0–10) to different performances. Thus, all those maps formerly built are reclassified on the basis of 10 intervals, where 0 represents the lowest score and 1 or 10 the best one. This step is essential for two reasons:

- intervals have to be refined according to the object under analysis;
- a-dimensional values allow maps comparison and aggregation.

The standardization has been carried out by drawing value functions through the Evalue techniques (Demetriou et al. 2012). By the value functions, it's possible to check, for example, the respondent's willingness to walk or bike a certain distance to

[2]https://www.agenziademanio.it/export/sites/demanio/download/schedeapprofondimento/REPORT-SHORT_CAMMINI-E-PERCORSI-da-5.0.pdf.

Table 1 Set of criteria

Criteria	Sub-criteria	Description
Natural resources	Parks and protected areas	The criterion considers protected areas aimed to preserve ecosystems at national, regional and supra-municipal levels
	Natural areas	The criterion considers protected areas that are part of the European ecological network Natura 2000
Historical-cultural resources	Cultural heritage	The criterion considers the architectural assets that represent the historical and cultural heritage
	Landscape	The criterion considers archaeological sites, i.e., places where traces of human activities from the past are preserved, as well as the panoramic points present in the territory and the navigable waterfronts
Socio-economic resources	Hospitality	The criterion considers the accommodations network, consisting of room rentals, farmhouses, hotels, B&B, camping, country houses, guest houses and hostels
	High quality wine-yards	The criterion considers the areas of fine vineyards, i.e., the areas devoted to the cultivation and production of DOC, DOCG and IGP wines
	Connectivity	The criterion considers the connectivity of the routes with other cycle paths and hiking networks
	Interchange nodes	The criterion considers railway stations and riverside ports, which allow connections with other transport systems

reach a specific place. An interview has been administered to one expert for each criterion (Natural resources, Historical-Cultural resources, Socio-Economic resources) according to a procedure divided into the following four steps:

1. selection of a range and type of curve;
2. values assignment;
3. revision of the curve;
4. consistency control.

In the first step, the range to be used to standardize each sub-criterion is identified and the expert is asked to define the type of curve, i.e., if it will therefore have a rising, falling, constant, Gaussian trend. Subsequently, it is possible to transform the performances in a-dimensional values considering the defined range, by asking to the expert to specify a maximum and minimum value and then to proceed with the intermediate ones. During the third and the fourth steps, the function is reviewed together with the expert. Table 2 presents the value functions defined for each sub-criterion and provides a description for each them.

The Design phase involves also the criteria weights elicitation, a crucial step in the definition of the influence of the criteria and sub-criteria during the decision process. The aim of this step is to assign a different importance to the elements involved in the problem since, according also to a multidimensional perspective, they can play a different role in achieving the final purpose. Weights assignment makes judgments explicit. As suggested by Riabacke et al. (2012), weights elicitation methods should be carefully selected according to the decision problem and the stakeholders involved.

3.1.3 Choice Phase

Once the maps have been standardized (Criterion Maps) and weights assigned, it is possible to proceed with the aggregation phase which implies the choice between compensatory or non-compensatory methodologies. In the first case, values are aggregated and a negative performance according to one criterion is compensated by a good performance for another one; in the second case thresholds are defined, and if the performance does not meet the requirements, the alternative under evaluation cannot be considered as acceptable. One of the most common compensatory methods is the Weighted Linear Combination, WLC:

$$Aj = \sum WiXi \tag{1}$$

where:

Aj alternative represents the suitability of the j-th pixel in the final map;
Wi is the normalized weight of i-th criterion;
Xi is the standardized score (Malczewski 2000).

Since each criterion and dimension is defined by pixels, in the formula the suitability is related to each pixel.

The result of the aggregation procedure is the Suitability Map that represents the territorial attractiveness of placed crossed by historical-religious itineraries and cycling/walking trails.

One of the strengths of the SMCA is the possibility to make all the evaluation phases explicit and clear, starting from the intermediate maps and aggregations (Partial maps) to the final Overall map. By such an evaluation process, decisions about where and how to preserve cultural heritage are robust, as they have been based on a deep investigation of territorial potentials.

Table 2 Sub-criteria value functions

Criteria	Sub-criteria	Standardization	Value function
Natural resources	Parks and protected areas	Euclidean distance Distances ≤5 km are standardized to 10 Distances > 5 km are standardized to 0	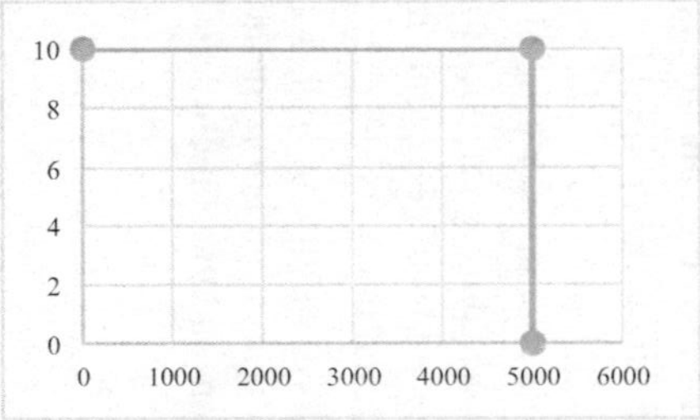
	Natural areas	Euclidean distance Distances ≤5 km are standardized to 10 Distances > 5 km are standardized to 0	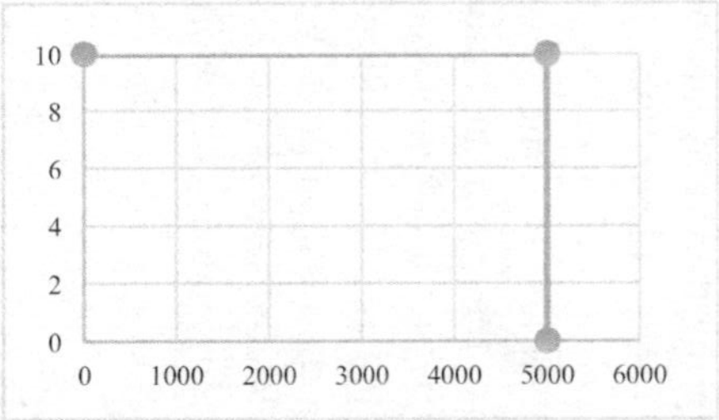

(continued)

Table 2 (continued)

Criteria	Sub-criteria	Standardization	Value function
Historical-cultural resources	Cultural heritage	Point density Density ≥36 public assets are standardized to 10. Density between 36 and the minimum are standardized according to linear score (the higher Density, the higher score)	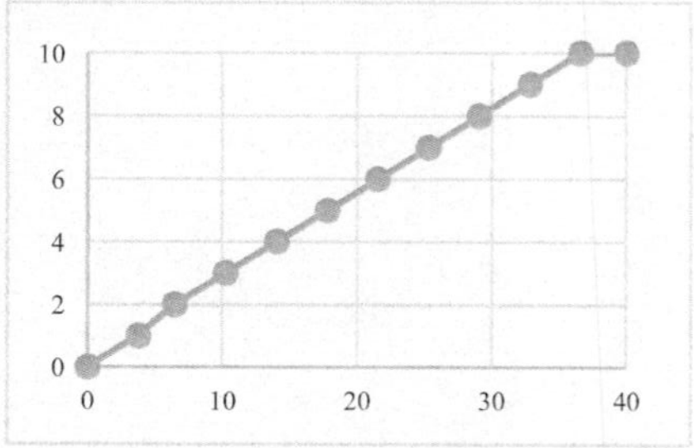
	Landscape	Archaeological heritage and vantage points Point density Density ≥0,8 public assets are standardized to 10. Density between 0,8 and the minimum are standardized according to linear score (the higher Density, the higher score)	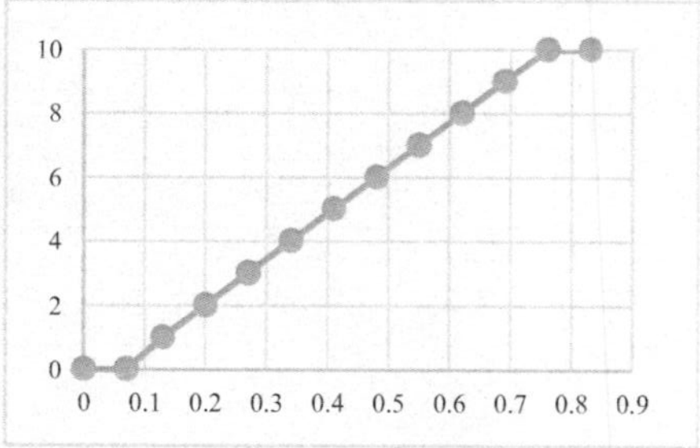

(continued)

Table 2 (continued)

Criteria	Sub-criteria	Standardization	Value function
	Landscape	Fairways Euclidean distance Distances ≤3, 8 km are standardized to 10. Distances between 3, 8 km and the max are standardized according to the linear score (the lower the Distances, the higher the score)	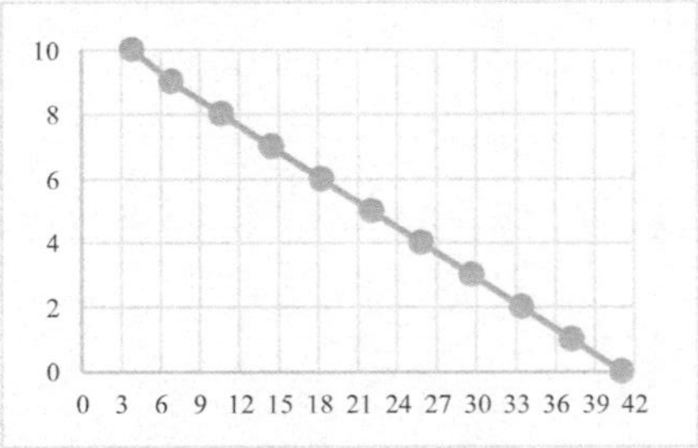
Socio-economic resources	Hospitality	Euclidean distance Distances ≤1 km are standardized to 10. Distances between 1 km and 5 km are standardized to 5. Distances between 5 and 10 km are standardized according to the linear score (the lower the distances, the higher the score)	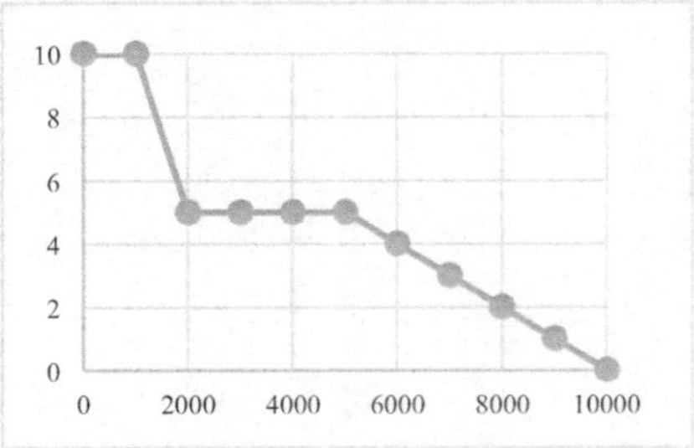

(continued)

Table 2 (continued)

Criteria	Sub-criteria	Standardization	Value function
	High quality wine-yards	Euclidean distance Distances ≤1 km are standardized to 10. Distances ≥1 km are standardized to 0	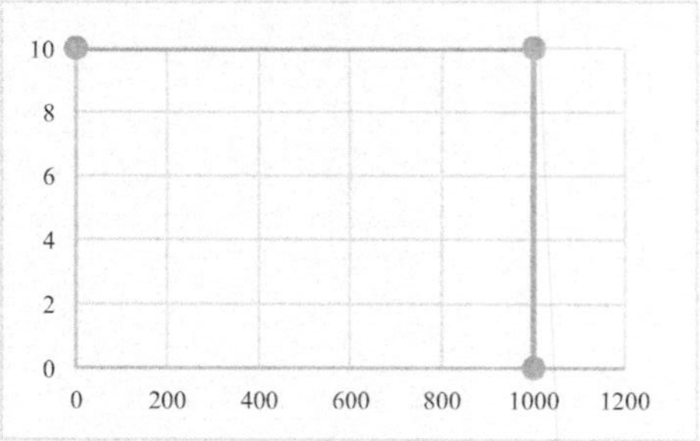
	Connectivity	Euclidean distance Distances ≤3 km are standardized to 10. Distances ≥3 km are standardized to 0	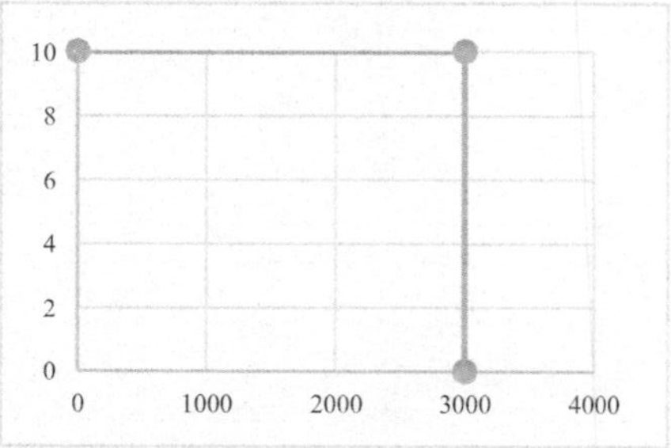

(continued)

Table 2 (continued)

Criteria	Sub-criteria	Standardization	Value function
	Interchange nodes	Euclidean distance Distances ≤3 km are standardized to 10. Distances between 3 km and 5 km are standardized to 9. Distances ≥5 km are standardized to 0	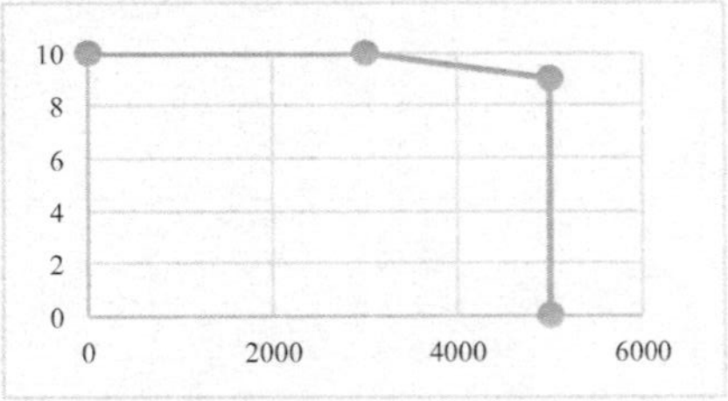

3.2 Territorial Vulnerability Assessment

In parallel with the analysis of the territorial attractiveness aimed at assessing strengths of the areas crossed by historical-religious itineraries and cycling/walking trails, it has been developed the analysis of the territorial vulnerability aimed at assessing the weaknesses. Considering the full range of aspects relevant to the decision problem improves the quality of the final decision (Bottero and Ferretti 2010). In fact, when complex decision has to be done, taking into account also negative issues allows to detect potential criticalities and risks that could have direct effect on the project. Vulnerability could be described both as the exposure of elements to adverse impacts (Alexander 2012) and as the measure of the damage (Blaikie et al. 1994) or as a combination of the exposure and the resistance (Villegas-González et al. 2017; Alexander 2012). In this context the socio-economic, socio-cultural and the natural context are going to be evaluated both considering negative impacts on the territory and their measure. Below the characteristics analysed are better explained.

- Environmental Risk analysis:
  - Flood and Hydrological risk: given the presence of rivers, it is crucial to know the areas where the hydrogeological or flood risk could occur. The determination of the risk class derives from the assessment of the hazards, connected to the different types of instability, socio-economic and infrastructural vulnerability of the areas potentially subjected to damage as a result of the occurrence of instability phenomena. This assessment procedure allows to assign four risk classes to the municipalities.
  - Seismic Risk: equally important is the knowledge of seismic risk and of the damages to which the territory could be subjected, in order to understand how to reduce them. Italian municipalities are classified according to the probability that the territory will be affected in a certain period of time (usually 50 years) by an event that exceeds a certain threshold of intensity or magnitude.
- Demographic Vulnerability:
  - population density,
  - age of the population
  - education.
- Economic Vulnerability: the analysis of the municipalities from an economic point of view allows to understand which sectors are most developed and therefore offer employment to the population. It is also important to understand whether a municipality can guarantee work for most residents, or whether they are forced to become commuters.

In order to result in a territorial vulnerability value, data previously analysed are processed and aggregated by considering the risk and the socio-economic perspective, in order to understand which assets are located in a more fragile context and could be at risk.

Fig. 4 Priority matrix

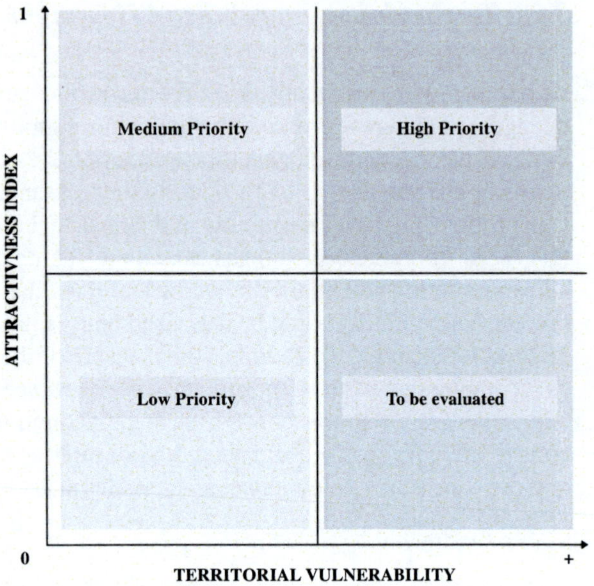

3.3 The Priority Matrix

For what concerns the third phase, the results of the two previous steps have been combined into the Priority matrix, based on the Attractiveness index and on the Social and Territorial Vulnerability value (Fig. 4). The first variable on the vertical axis represents the Attractiveness index, the second, on the horizontal axis, represents the qualitative score assigned to the Social and Territorial Vulnerability. How it emerges from Fig. 2, the matrix allows to decide reuse interventions and to understand the level of priority of assign to each cultural heritage located nearby historical-religious itineraries and cycling/walking trails. In fact, properties that require a priority renovation intervention, or anyway that deserve to be re-functionalised are the ones located in a territory with a high level of attractiveness and vulnerability. Those located in a territory with a high level of attractiveness and low level of vulnerability obtain a medium priority, low priority for those ones with a low level of attractiveness and a low level of vulnerability, while assets positioned in areas with low level of attractiveness and high level of vulnerability deserve to be better investigated in order to evaluate their potentials with respect to weaknesses and threats of the territorial context where they are located.

The subsequent phase concerns the investigation of the intrinsic features of public assets to understand their current conditions and to proceed to program. By this kind of multi-step procedure, decision-makers will be able to consider cultural heritage's potentials and criticalities together with the territorial attractiveness and vulnerability. In order to activate spread territorial catalytic effects, it could be of strategic importance to start with the reuse of cultural heritage located in attractive and vulnerable territories, as this category of assets is at risk and their reuse can benefit of the territorial context strengths and opportunities.

4 Conclusions

The valorisation of cultural heritage is a complex decision problem given the instance of taking into account multiple territorial values and constraints as well as the quality of the assets, according to a sustainable balance among environmental and socio-cultural context and economic development. Moreover, these values are site specific and the analysis of the context plays a crucial role (Oppio and Bottero 2018).

The definition of a multi-methodological approach, based on the spatial analysis of territorial strengths and criticalities, seems to be a promising approach both to be aware about the territorial potentials and constraints and to assign a priority order to the interventions on cultural heritage. According to the steps previously described, the evaluation methodology could have a twofold purpose, predictive and supportive. In the first case, by taking into consideration at the same time multiple criteria, the evaluation framework allows to check whether the existing as well as the designed routes cross or approach areas with a high territorial attractiveness index, thus supporting new slow paths design or new services location. With respect to the supportive role evaluation can play, ranking cultural heritage reuse interventions according to the results of the Priority matrix allows municipalities to decide where to address efforts for making these interventions feasible, better addressing the limited resources available and improving the transparency of the choices on cultural heritage toward a common interest.

In order to test the robustness of the proposed multi-methodological approach, it's crucial to test it on a real case study included in the Agenzia del Demanio initiative with a special attention to the following main issues: (i) how to map and evaluate intangible values; (ii) how and in which step to take into consideration different categories of stakeholders; (iii) how to properly consider the line perspective into evaluation. These points need to be deeply investigated to make the evaluation process effective and appropriate to the specific features of the slow mobility lines.

References

Alexander D (2012) Models of social vulnerability to disasters. RCCS Ann Rev. A selection from the Portuguese journal Revista Crítica de Ciências Sociais (4)

Blaikie P, Cannon T, Davis I, Wisner B (1994) At risk – natural hazards, people's vulnerability and disasters. London

Bottero M, Ferretti V (2010) An analytic network process-based approach for location problems: the case of a new waste incinerator plant in the Province of Torino (Italy). J Multi-Criteria Decis Anal 17(3–4):63–84

Bottero M, Datola G, Monaco R (2019) Fuzzy cognitive maps: a dynamic approach for urban regeneration processes evaluation. Valori e Valutazioni 23:77–90

Bowitz E, Ibenholt K (2009) Economic impacts of cultural heritage, Research and perspectives. J Cult Herit 10(1):1–8

Burrough PA, McDonnell R, McDonnell RA, Lloyd CD (2015) Principles of geographical information systems. Oxford university press

Camagni R (2009) Territorial capital and regional development. In: Capello R, Nijkamp P (eds) Handbook of regional growth and development theories; Černá A, Černý J, Malucelli F, Nonato M, Polena L, Giovannini A (2014) Designing optimal routes for cycle-tourists. Transp Res Procedia 3:856–865

Chan K, Satterfield T, Goldstein J (2012) Rethinking ecosystem services to better address and navigate cultural values. Ecol Econ 74:8–18

Damart S, Roy B (2009) The uses of cost–benefit analysis in public transportation decision-making in France. Transp Policy 16(4):200–212

Dell'Ovo M, Capolongo S, Oppio A (2018) Combining spatial analysis with MCDA for the siting of healthcare facilities. Land Use Policy 76:634–644

Demetriou D, Stillwell J, See L (2012) An integrated planning and decision support system (IPDSS) for land consolidation: theoretical framework and application of the land-redistribution modules. Environ Plan 39(4):609–628

Diti I, Torreggiani D, Tassinari P (2015) Rural landscape and cultural routes: a multicriteria spatial classification method tested on an Italian case study. J Agricult Eng 46(1):23–29

Ferretti V (2011) A multicriteria spatial decision support system development for siting a landfill in the province of Torino (Italy). J Multi-Criteria Decis Anal 18(5–6):231–252

Ferretti V (2012) Verso la valutazione integrata di scenari strategici in ambito spaziale. I modelli MC-SDSS, Celid, Torino

Giaoutzi M, Nijkamp P (2006) Tourism and regional development: new pathways. Ashgate, London

Hoang H, Truong Q, Nguyen A, Hens L (2018) Multicriteria evaluation of tourism potential in the central highlands of Vietnam: combining geographic information system (GIS), analytic hierarchy process (AHP) and principal component analysis (PCA). Sustainability 10(9):3097

ICOMOS (2018) Annual report of the international council on monuments and sites

Ingold T (2015) The life of lines. Routledge

Keeney RL (1992) Value-focused thinking. In: A path to creative decision making. Harvard University Press, Cambridge

Leriche F, Daviet S (2010) Cultural economy: an opportunity to boost employment and regional development? Reg Stud 44(7):807–811

Malczewski J (2000) On the use of weighted linear combination method in GIS: common and best Practice approaches. Trans GIS 4(1):5–22

Moscarelli R, Pileri P, Giacomel A (2017) Regenerating small and medium sized stations in Italian inland areas by the opportunity of the cycle tourism, as territorial infrastructure. City Territ Archit 4(1):13

Munda G (2004) Social multi-criteria evaluation: methodological foundations and operational consequences. Eur J Oper Res 158(3):662–677

OECD (2001) Territorial outlook. OECD, Paris

Oppido S, Ragozino S, Icolari D, Micheletti S (2018) Landscape as driver to build regeneration strategies in inner areas. A critical literature review. In: International symposium on new metropolitan perspectives. Springer, Cham, pp 615–624

Oppio A, Bottero M (2018) A strategic management based on multicriteria decision analysis: an application for the Alpine regions. Int J Multicrit Decis Making 7(3–4):236–262

Oppio A, Bottero M, Ferretti V, Pracchi V, Fratesi U (2015) Giving space to multicriteria analysis for complex cultural heritage systems: the case of the castles in Valle D'Aosta region, Italy. J Cult Herit 16(6):779–789

Pearce D, Nash C (1981) The social appraisal of projects. In: A text in cost-benefit analysis. MacMillan Press, London

Quattrone M, Tomaselli G, Riguccio L, Russo P (2017) Assessment of the territorial suitability for the creation of the greenways networks: methodological application in the Sicilian landscape context. J Agricult Eng 48(4):209–222

Riabacke M, Danielson M, Ekenberg L (2012) State-of-the-art prescriptive criteria weight elicitation. Adv Decis Sci 2012:1–24

Rodríguez-Pose A, Vilalta-Bufí M (2005) Education, migration, and job satisfaction: the regional returns of human capital in the EU. J Econ Geogr 5(5):545–566

Roodhouse S (2007) Cultural quarters: principles and practice. Intellect, Bristol

Roy B (1996) Multicriteria methodology for decision aiding, vol 12. Kluwer Academic Publisher, Dordrecht, Of Nonconvex Optimization and its Applications

Roy B (1999) Decision-aiding today: What should we expect?. In: Multicriteria decision making. Springer, Boston, MA, pp 1–35

Roy B (2005) Paradigms and challenges. In: Multiple criteria decision analysis: state of the art surveys. Springer, New York, NY, pp 3–24

Saarikoski H, Mustajoki J, Barton DN, Geneletti D, Langemeyer J, Gomez-Baggethun E et al (2016) Multi-criteria decision analysis and cost-benefit analysis: comparing alternative frameworks for integrated valuation of ecosystem services. Ecosyst Serv 22:238–249

Simon HA (1960) The new science of management decision

Simon HA (1979) Information processing models of cognition. Annu Rev Psychol 30(1):363–396

Torrieri F, Batà A (2017) Spatial multi-criteria decision support system and strategic environmental assessment: a case study. Buildings 7(4):96

Villegas-González PA, Ramos-Cañón AM, González-Méndez M, González-Salazar RE, De Plaza-Solórzano JS (2017) Territorial vulnerability assessment frame in Colombia: disaster risk management. Int J Disaster Risk Reduct 21:384–395

A Line Born for Sustainable Development: The Case Study of VENTO

Slow Travel Project for Enhancing Territories: Motivations and Directions

Paolo Pileri

Abstract Every project should follow an idea, a theoretical and intentional framework orienting the sense of its practical actions. Nevertheless, in many cases the project comes before then the motivations. This can be a mistake, especially in cases of regional projects affecting policies, people and territory. Giving priority to the idea makes the project more effective and people and institutional stakeholders involved more aware. This is the reason why the project VENTO, the proposal of a long cycle-tourist path in Italy, is born firstly by elaborating its motivations and objectives and, later, by designing the bike track. VENTO has the precise aim of regenerating marginal areas thanks to the introduction of the paradigm of the slow travel: it is a model of sustainable development with the shape of a cycle path. This paper analyses motivations and goals of the project and explains in which sense VENTO can be considered as a project of territory, as a way to increase local green jobs, as a political manifesto, as a new deal for investing in infrastructures, as a proposal for travelling and not for spending just few hours, as a vernacular opportunity, as a anti-fragility policy. These ideas have sustained the process of realization of the project.

1 A Cycling Tourism Project is First a Project of Territory and Cooperation Between Municipalities

VENTO is a cycling backbone route that is 700 km long and will connect VENezia and TOrino (Venice and Turin in english) running next to the Po River (Fig. 1).

For European readers, a cycling path is not news. Germany, Austria, France, The Netherlands, Great Britain and many other countries have had a cycling path network and have known the results that are generated for years. In Italy and other countries in the Mediterranean area there is almost nothing. The little cycling we have is

P. Pileri (✉)
Politecnico di Milano, Italy
e-mail: paolo.pileri@polimi.it

P. Pileri and R. Moscarelli (eds.), *Cycling & Walking for Regional Development*, Research for Development, https://doi.org/10.1007/978-3-030-44003-9_10

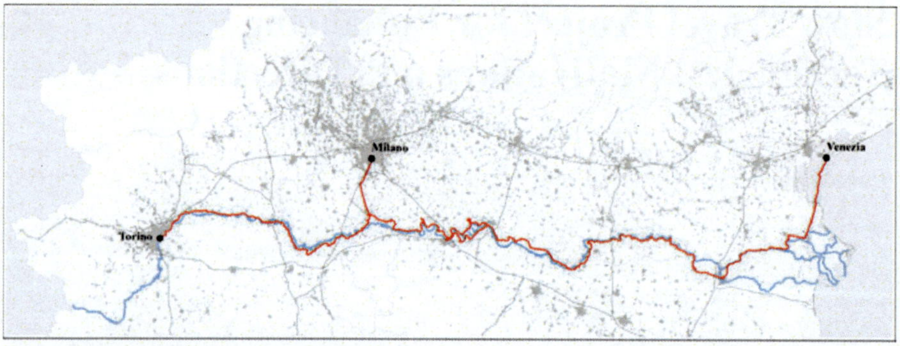

Fig. 1 The cycle path VENTO in red. *Source* VENTO Research Group, 2019

mostly urban. In Italy, there are few touristic infrastructures: some hundred kilometres in Trentino Alto-Adige and some short stretches of a few tens of kilometres here and there, often not continuous and with constructive and security criteria that vary from stretch to stretch. The result is totally underdeveloped cycling tourism to the point where it is not included in any annual statistic of the Italian National Tourism Agency (ENIT.it). Between 2008 and 2010 our research group at the Politecnico di Milano studied the great cycling routes of central Europe to understand the reasons behind this public success. With a background as territorial planners, our focus was on the positive impacts that these cycling tourism flows generated in the territories. When planning the VENTO cycling route, we were convinced that a more territorial outlook rather than cycling or sports-oriented or recreational was decisive in the design of the project. This great cycling route, whose length had not been even thought of before—700 km—became the proposal for a new model of development and not (only) the study of an alternative way of moving and travelling. A development model that includes advantages for the territories that are crossed (employment, economy, territorial protection, depopulation control, …), and advantages for those who travel on slow routes (learning, landscape appreciation, culture, health, environmental defence, …). In this sense, we can say that the project of a cycling route must first of all be seen as a territorial project, and then as the realization of a cycling infrastructure with all of its technical features, including paving, curbs, signs, rest areas, etc. And even after that, as an activity to promote tourism. If then, like in the case of VENTO, the cycling route crosses marginal areas of the country, the motivation that the project is based on becomes even more ambitious and urgent, as it is a matter of regenerating fragile territories. The road infrastructure becomes a simple tool without which, however, we do not consider it is possible to start a concrete regeneration as without a cycling route those slow tourism flows that are capable of distributing economy are not triggered. A cycling route like VENTO must be able to be travelled by 300–500,000 cycling tourists a year, against the few thousands of today, where there is no well-made cycling route between Venice and Turin. Only with these numbers can sufficient finances to generate entrepreneurship and new jobs be triggered. The long research work carried out between 2009 and 2011 made us discover that the great European cycling routes were very frequented and their tourists were everyday people, not cycling experts, otherwise there would be fewer of them. Therefore, the technical project of the cycling route must keep in mind that its "passengers" are inexperienced, they are children, they are disabled, they are older people who need comfort, ease and safety. This is why VENTO is thought out as a cycling and pedestrian track that is not open to cars or motorbikes for 90% of its length. The paving chosen for the VENTO project is smooth, the roadway is wide, continuity is ensured. This is what guarantees safety and makes the number of participants increase.

2 Designing a Vision For a New Slowness Geography to Live

Along its 700 km, VENTO touches 150 municipalities, 4 regions, 13 provinces, 2 river agencies, 47 protected areas, a dozen UNESCO sites, a dozen museums, various water reclamation consortiums, 1 million inhabitants, 3000 commercial activities and more than 10,000 farms. Many realities that have not been able to converge in a project that is common, as they lack the impulse towards cooperation and a vision in which they can recognize themselves. With VENTO, we tried to design an ideal but also concrete vision, as a cycling route is something that can be really feasible, on which an imaginary of possible future can be deposited. Our research group at the Politecnico di Milano drew up VENTO's feasibility study in 2010, demonstrating that what was only a trace on paper could concretely become a true infrastructure on which to deposit the hopes for local development. To make local administrators and citizens understand the importance of cycling tourism as a cooperative and territorial project, in 2013 we started cycling between the squares of villages and towns in the Pianura Padana, becoming cycling ambassadors ourselves of the VENTO project and the advantages of its development model (www.VENTObicitour.it) (Fig. 2).

It was therefore a natural progression to become interlocutors with the National Government asking to open a technical committee to start the realization of a proper cycling route along the Po River. This happened in 2016 with the initiation, on the part of the Ministry for Public Infrastructure, of the Sistema Nazionale della Ciclabilità Turistica (the national system of cycling tourism) with their own financing. Finally, Italy started to have their own budget to create great bicycling routes and not simply urban cycling paths. The feasibility projects then started in 2018.

Fig. 2 Group of cyclists participating at VENTO Bici Tour 2018. *Source* VENTO Research Group, 2018

3 Investing in Slow Infrastructures to Generate Green Jobs

VENTO introduced the concept of cycling backbone route, a cycling path that becomes the main axis for a possible collection of local connections and that then connects with other axes. The model is inspired by the German long-distance cycling networks and the other great European cycling routes like the one along the Danube, the Drau, the Inn and so on, the result of a strong regional and national direction (Pileri et al. 2015). The great European cycling tourism is not a product of chance, but the fruit of a long and meticulous project both territorially and infra-structurally with which vision and scale of action have changed and where the pleasure of travelling for hundreds of kilometres has substituted the urgency of movement. This is what attracts thousands of visitors for more and more days. Germany alone created, in just 30 years, a long-distance pedestrian and cycle pathway, that is continuous, easy, accessible to all, over 45,000 kms long and where every year a cycling tourism flow of 175 million holiday days are spent on bikes, totalling a turnover of just under 9 billion euro (Pileri et al. 2015). The German cycling tourism network gives full-time employment to over 200.000 people (Blondiau and van Zeebroeck 2014), ten times as many as statistics show for Italy. And it's precisely the objective of generating employment that has fascinated VENTO, making it one of the pillars of its design proposal. European cycling routes manage to combine high employment benefits (in Europe 5 employed people per km in the best cases) in view of low public investments (2–300,000 euro/km, a hundred times less than the cost of a kilometre of highway construction) and with a higher guarantee of distribution of social and economic benefits. This all proved to be a strong motivation for VENTO to propose its infrastructural and territorial vision firmly. Within cycling tourism, as in walkways, there are various types of employment, decent and appropriate to young people: restauranteurs, hotel and lodging managers, local producers, farmers, cultural guides, personal service providers, carriers, mechanics, healthcare providers, architects, urban planners, engineers, craftsmen and women, etc.

4 Slow Travel Is Free and Hardly Accompanied: Good Infrastructure Is More Crucial Than Marketing

The study of cycling tourism in Germany, well documented by ministerial and association's reports in the country, revealed that 84–88% of cycling tourists organize their own slow trip autonomously, so they do not use the services of tour operators (ADFC 2019). This circumstance can probably be explained by the fact that the consistency, the security and the capillarity of the cycling network are such as can favour the free flow, increasing the number of visitors. Successful cycling tourism in that country seems more in debt to the infrastructure rather than to touristic marketing and tour operators. The lesson we can take away from this is to give priority to the necessary structural and infrastructural investments and only after that to launch touristic marketing. Inverting these factors is risky because it attracts potential tourists who then do not find the cycling route in the conditions they expect. Thanks to territorial marketing, the entrepreneurship of tour operators has some undoubted advantages and they can even, in some way, compensate for the lack of safe cycling routes, accompanying travellers step-by-step, but in the long term the risk is to keep the number of visitors low, damaging the territories and, consequently, also the tour operators. VENTO has chosen the philosophy of slow and free travel, based on a safe, long and continuous cycling route where travellers can stop where they prefer and where their sensitivity suggests. Cycling travellers eat wherever they want; they visit what is to be seen based on their preferences; they meet who they are attracted to, taste foods, relax, choose other complementary activities based on their needs and tastes. Each person colours its journey within the contours suggested by the track he has chosen. The accompanied tourist, on the other hand and unknowingly, risks to give himself up to taste the freedom of autonomous discovery that is crucial for the experience of slow travelling. We cannot therefore do without cycling infrastructure. Tour operators remain important actors, but their presence cannot become the alibi behind which the public decision-maker comes to the decision not to invest in infrastructures leaving them to fill the gaps of a riddle network. By doing this, slow tourism risks to agonize before being born because of a lack of numbers. Doing so, the risk is to not activate those regenerative effects that are widespread and strategic for territories as the focus of tour operators lies only where they see more profit for themselves and not for the territory. Cycling and walking paths can actually create public leverage, which is important precisely to give a chance of release to the most marginal and fragile areas, where there are currently no tourism economies. But this political choice can only be carried out by a public subject.

5 Long Cycling Travel and Short Cycling Trips. Is It the Same Planning Approach?

Since the beginning, VENTO's ideal cycling tourist is a person who is not necessarily an expert and who carries out a cycling tour of more than one night: the best is at least three nights according to the German's definition of cycling tourist (ADFC 2019). It technically means creating walking and cycling tracks between 200 and 300 km long (hypothesizing an average daily distance between 50 and 75 km, it means three-four travel days). This ambitious combination of user/infrastructure better guarantees an average expenditure that is vital for the territories that are crossed: 64 €/diem per cycling tourist (BMWI 2009). Bike-hiking, or the cycling day trips on short distances from 10 to 30 to 50 km, not far from home or from hotels, are on the other hand tied to a more modest daily expenditure, 16 €/diem per person, causing a potential income which is much lower for the territories that are crossed. Between bike-hikers and tourists who occasionally use their bicycles, VENTO has chosen cycling tourists, as they spend more days on the route (and not in one place) and because it enables the discovery of places in an osmotic way (Christin 2019). Thus, the traveller continuously learns and discovers different things along the way. And the territory itself benefits the most, increasing financial income and their distribution. In Germany every year this type of cycling traveller generates an income that ranges between 4 and 4.5 billion euro that are deposited in the internal areas that are crossed by the great cycling routes. Oxygen for sustainability and for local companies.

6 A Vernacular Line, Not Just a Commercial Opportunity

The VENTO track took inspiration from the Eurovelo 8 proposal, but it differentiates itself for having favoured immersive routes in ordinary landscapes and in the vernacular context. VENTO's track encourages the traveller to discover the value of the variety of lesser known landscapes, local foods, folk stories. We are convinced that in this fine and almost forgotten weave lies the authenticity that is still capable of narrating the territories according to a non-obvious and very sophisticated spontaneity. Furthermore, according to Rudolph Christin, slow travel has precisely the power to re-localize the body and awaken the attention to the world, "which means finding the spirit of adventure again even in places that are at first glance insignificant" (Christin 2019, p. 124). The vernacular side of the territory is a precious material for a project such as VENTO and therefore must be safeguarded even when project solutions are being decided on. For example, not foreseeing rest areas that are equipped with food courts along the walking and cycling route can seem a failure to pay attention to the traveller, but it is instead a choice that defends the local dimension of the territory, as it invites the traveller to go to the local villages, to use the local inns or trattorias, which will in turn experience a financial benefit because of them. Even guaranteeing the survival of inns by helping them to remain true to themselves and faithful to their traditions is part of a cycling tourism project with a cultural claim. If the cycling tourism project wants to take care of the local sphere, it must distance itself from certain touristic-commercial formats that have nothing to do with slow travel.

7 An Enchanting Narrative, Not Only Informative Signals

Road signals are obligatory on all cycling routes in Europe. But the predisposition to curiosity of the slow travellers cannot be content with obligation or directions signs, however, useful they may be. There is a need to create some communication tactics that are able to pleasantly and intelligently reveal what happens around the line where you walk and cycle. We have called all of this a "narrative project" that gives new voice to landscapes that are encountered. If today many people are increasingly less aware of the history and the stories of places outside of cities, slow travel is a great opportunity to reconnect them to landscapes. But landscapes need to go back to talking about themselves by developing new languages. For example, by using symbols, key words, icons and anything that can trigger a simple curiosity that can light up the desire in passers-by. A cycling route like VENTO can be a real narrative line where no passer-by will be indifferent to the places in which they venture. This is why we believe that a narrative project for the entire cycling path and its surroundings should always be ready. Figurative art, in its various expressions, from street art to design of objects to infographics to other forms, is a possible field to experiment new communicative codes that are capable of establishing simple but never trite contact with those it passes by. Beyond graphic solutions, the VENTO experience teaches us that a line project without a great narrative project doesn't work. The narration we propose is something quite different from territorial marketing, whose primary aim is to make travellers spend as much money as it's possible and not firstly to take care about their cultural development (Fig. 3).

Fig. 3 From road signals to narrative graphics and writings to tell slow travellers the landscape's stories. *Source* Pileri et al. (2018) renderings from a proposal made by VENTO lab to a Municipality in Veneto, Italy

Fig. 3 (continued)

8 Let's Mend Beauty: A Civic Manifesto Hanging From a Thread

Among the main possible interpretations, designing a cycling route means also committing to a sort of ambitious plan of action that consists in mending through people and threads a territory that is otherwise broken down in thousands and thousands of administrative fragments, made of municipalities competing one against the other. Designing a cycling route with the awareness that it is a thread that sutures the wounds of a torn territory lays down a strong change in perspective and an increase in civic responsibility in those who design. At the end, it's not a missing infrastructure that is being added, but a precise educational project as much for the territory as for those who will use it. It will be the walkers and cyclists along this track thus designed themselves who, step-by-step, with their own eyes, their intelligence, will be able to continually mend together the patches that have comes loose in the territory. We all mend, involved in a project of regeneration of care, even before a regeneration of places. Care for what? For beauty. This is the true common factor of what we patiently meet along the VENTO line (Img Ricuciamo la bellezza). This is the great reason for which it is worth investing in a great opus like the 700 km cycling route. A beauty made of an infinity of material and immaterial heritages dispersed here and there. Of historical sedimentation and folk stories. Of nature and culture. Of goodness and tradition. A beauty that to save itself needs the eyes of those who can admire it, otherwise it disappears. Eyes that in the internal and marginal areas of landscapes outside cities are increasingly rare or go by too quickly. Without eyes and without intelligence, every beauty fades and goes extinct, just as a plant species goes extinct. The hope we place in slow projects of paths, cycling paths, cycling and walking routes, walkways, water routes, bridleways is that they can bring back beauty to everyone's gaze before it fades away. Eyes to train. Intelligences to be educated. Eyes and intelligence that thanks to slowness can take care of landscapes. Besides, the root of the Italian word for caring—cura—has a -ku root that in ancient times meant observing, as if to say that nothing takes care of itself, or better, everything can be cared for if it is observed, auscultated, analysed before. Step after step, ride after ride, those who travel slowly end up taking care of those places, which only this way will never cease to exist and will be able to hope that someone, having got to know them, will defend them. Let's mend beauty becomes a civil manifesto for linear and slow projects such as VENTO (Fig. 4).

Fig. 4 Learning heritage walking or cycling it's an opportunity offered by a slow line if designed in a such way. *Source* VENTO Research Group, 2018

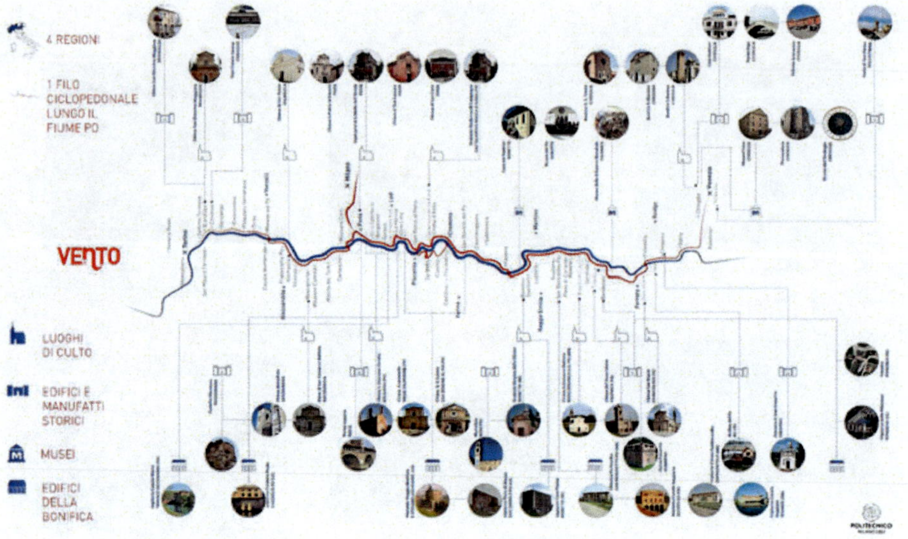

Fig. 4 (continued)

References

AA.VV (2009) Grundlagenuntersuchung Fahrradtourismus in Deutschland. Bundesministerium für Wirtschaft und Technologie, BMWi, Berlin

ADFC (2019) Travelbike bicycle travel analysis—Summary report. https://www.adfc.de/fileadmin/user_upload/Expertenbereich/Touristik_und_Hotellerie/Radreiseanalyse/Downloads/Datenblatt_Radreiseanalyse2019_en_-_neu.pdf. Accessed 10 Februaary 2020

Blondiau T, Van Zeebroec B (2014) Cycling works. ECF, Brusselles

Christin R (2019) Turismo di massa e usura del mondo. Eleuthera, Milano

Pileri P, Giacomel A, Giudici D (2015) VENTO. La rivoluzione leggera a colpi di pedale e paesaggio. Corraini editore, Mantova

Pileri P et al (2018) Ciclabili e cammini per narrare territori, Ediciclo, Portogruaro (VE)

Design, Public Engagement and Communication: Reframing Methodology

Alessandro Giacomel, Diana Giudici, and Camilla Munno

Abstract VENTO is a territorial project threading its way through the cycling–pedestrian infrastructure that will run along the banks of the river Po from VENice to Turin [TOrino]. The project has been conceived at Politecnico di Milano to propose a concrete and viable development option in response to the fragile situations of the inner areas around Italy's largest river valley. The project defines a new methodology to manage complex territorial governance plans, which concomitantly address both the vast and local scale. The scope of the paper is to describe the methodology developed by the research team by dividing the process into three phases: ideation, taking charge and implementation. The latter is still in progress. The plan, which has also been carried out with ongoing and constant cultural work targeting multiple subjects—spanning local, regional and central institutions, national and local associations, and citizens—was developed with a parallel communication project as important as the infrastructural project. Indeed, it is an intrinsic part of the methodology perfected with VENTO.

1 A Methodology Based on a Concrete Project: VENTO

Project VENTO was conceived at Politecnico di Milano, Department of Architecture and Urban Studies (hereinafter referred to as DAStU) in 2010 by a research team working on sustainable territorial planning with particular focus on the weakest and less favoured areas of the country. A multidisciplinary team that, with project VENTO, has played a key role in introducing to the public agenda the creation of cycling tourism infrastructures as a lever for the development of fragile territories. A

A. Giacomel · D. Giudici (✉) · C. Munno
Politecnico di Milano, Italy
e-mail: diana.giudici@polimi.it

A. Giacomel
e-mail: alessandro.giacomel@polimi.it

C. Munno
e-mail: camillamunno@gmail.com

P. Pileri and R. Moscarelli (eds.), *Cycling & Walking for Regional Development*, Research for Development, https://doi.org/10.1007/978-3-030-44003-9_11

project conceived without a client, carried out by an independent subject, foreign to professional rationale and, therefore, not subjected to deadlines. A project inspired by high ideals, without the limitations of either an assignment or a budget. VENTO has perfected a methodology to develop projects inspired by local resources to reassemble such resources in a territorial plan capable of mending the beauty of places, the pearls, into a chain, whose thread is the cycling-pedestrian infrastructure for tourism.

From infrastructural project to territorial project, "whose cultural paradigm lies in the concept of line. The line possesses a strength that mends, keeps together, reconstructs narration, offers cohesion, highlights the great river and its landscapes once again, gives consistency and continuity to small local enterprises, relaunches the many dispersed beauties by establishing a trend of interpretation without kneading the individual features, stories and traditions into a matt colour of single-minded, unemotional, marketing for tourism" (Alberti et al. 2019, p. 225): with this objective, the research team has perfected a methodology that can be replicated to offer development opportunities to fragile territories. The pages below analyse and describe the methodology proposed and refined with project VENTO, starting with guidelines for the reconstruction of complex decisional processes proposed by Dente (2011). The timeline of the decisional process, evaluation of actors and analysis of interactions and of causal relations through which the analysed process unfolds are interpretational keys adopted to examine and yield the research methodology used to develop project VENTO. A project launched ten years ago, in which today we can recognize three different phases of development from 2010 to date: (I) ideation; (II) taking charge; (III) implementation (in progress). This division into phases, by the authors, to describe the methodology proposed stems from the identification of factors that have determined relevant twists in the overall process. Contextual conditions are described for each phase, highlighting the issue the research team wanted to face, their intervention, the timeline of the main events characterizing that phase, the network of actors involved and the economic resources used/available. The communication project is given a place unto itself. It follows the analysis of phases as an element characterizing the methodology proposed, which is transversal to the three individual phases. The paper closes with a summary of the core ingredients of the methodology proposed by VENTO, reflecting on the cultural change triggered by the project. VENTO also contributes to change the approach to the territorial project.

VENTO is a complex project in terms of dimensions and for its original character in the national scene. It has required the creation of many tools and languages to dialogue with the many actors involved by the project. Research, project, collection of consent, identification of key actors who could be assigned its implementation and follow-up are the pillars of the methodology proposed, along with the communications project, which has accompanied every phase of the process, using new languages capable of addressing a public of heterogeneous subjects. Multiple languages have been adopted to convey form and substance to the essential cultural work, which has allowed the project to progress, introducing into the public agenda the theme of infrastructures for cycling tourism, and paving the way for that which has become the National System of Cycling Routes for Tourism[1] (hereinafter referred to as SNCT).

2 Phase I—Ideation

In 2010, unlike the situation across the Alps, projects for cycling tourism[2] and dedicated, continuous and safe long-distance cycling infrastructures didn't exist in Italy. There were multiple reasons that apparently differed in nature: bicycle trips were only taken by enthusiasts and accompanied groups; in the public agenda, the cycling theme was limited to the urban context, just as the technical regulations of the sector; cycling paths were almost only itineraries shared with cars and mutually disconnected due to administrative fragmentation; academic research considered "cycling tourism", "cycling planning", "fragile areas" and "participated planning" as mutually disjointed theme areas.

The disorganized picture, devoid of an idea of cycling tourism considered as a territorial project capable of combining all these aspects, was only waiting to be explored.

Project VENTO was launched in this problematic and fragmented context with a visionary idea that was also simple: link up Venice, Turin and Milan with a continuous and safe cycling-pedestrian route of 700 km[3] along Italy's longest river, the Po,[4] in a flatland rich in miscellaneous places that were beautiful, intense and packed with history (Fig. 1). Favourable orography with highly attractive city centres at the extreme ends: two suitable conditions to trigger a cycling tourism project, and to attempt to give new economic and occupational hope to fragile territories of inner areas[5] (Fig. 2) crossed by the cycling route.

---

[1] https://www.mit.gov.it/node/5383.

[2] With the exception of the Trento Provincial Administration, where the overall network of suburban cycling-pedestrian paths was ca. 200 km in 2011 and generated profits of ca. 86 million €.

[3] The length of the cycling route established by the 2012 feasibility study conducted by the Politecnico was 679 km. During the subsequent technical and economic feasibility planning phase (2018–2019), the length was increased to 700 km after several variants resulting from more detailed technical evaluations and some local requests that were accepted.

[4] For more than 50% of its length the VENTO cycling route runs along the main bank of the river Po. The bank is a hydraulic infrastructure with variable height from 2 m to over 10 m. A natural balcony on the surrounding landscape that could also ensure the safety of cyclists. Indeed, one of the project's goals is to make the bank's ridge accessible only to cyclists and to authorized vehicles.

[5] Inner Areas are scarcely accessible territories, which are far from the main urban centres and lacking in essential services (school, healthcare, railway transport) within a certain distance. In Italy, interest in these marginal sites was reawakened starting from 2014 with the National Strategy for Inner Areas (SNAI), which classifies ca. 60% of the national territory as 'inner area' (UVAL 2014).

Fig. 1 Landscapes along the route of the cycle path VENTO. *Source* Alessandro Giacomel, 2012

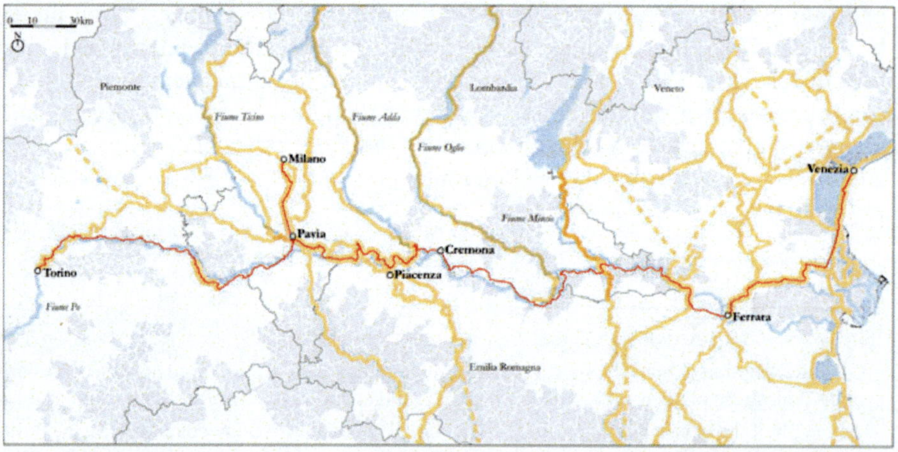

Fig. 2 VENTO (in red) and grey highlighted municipalities included in the classification of inner areas. *Source* the author, 2018

In this context, the idea was to initially focus on the structuring element, the cycling and pedestrian infrastructure that could create a territorial project. Hence, the first action was to arrange a feasibility study[6] of the VENTO cycling tourism

[6]The feasibility study is the first document that shows the actual possibility of implementing project VENTO. According to Italian technical regulations, the subsequent planning phases include: the

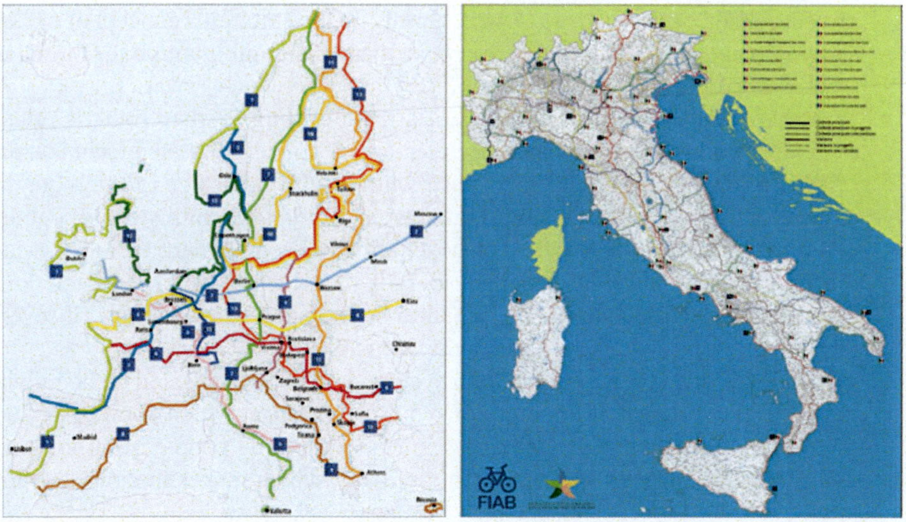

Fig. 3 Map of the Eurovelo network made up of 15 itineraries across Europe (left) and map of the 20 itineraries of the Bicitalia network (right). *Source* Eurovelo and Bicitalia

infrastructure, starting with the choice of the route. The initial idea was provided by the Bicitalia 2-Eurovelo 8 route[7] provided by the Italian Federation of Bicycle Friends (Fig. 3), which was later the focus of a campaign of more than 1000 km of surveys to study places and landscapes, and to test, verify and ascertain that the decision to cross at one point rather than at another was the best both in terms of landscape and of technical options. Just as "town planning is carried out on foot" as Bernardo Secchi often said in his conferences, cycling paths are designed on pedals, in the sense that work carried out at the desk does not suffice. It is, instead, essential to visit the territories, move at the speed of a bicycle, perceive the difficulties the traveller will encounter, check what he will see, etc. Hence, the survey phase was crucial for planning because only this approach could resolve doubts and find solutions that could not have been found otherwise, except "in the field" (Pileri et al. 2017). In parallel, work focused on defining some "design criteria", precisely five, with the intention of proposing a dedicated infrastructure that would overcome the precarious concept of itinerary by guaranteeing the following points:

- safety: by removing a large number of mixed tracks and discontinuities that might impair the safety of the cyclist, or even only his perception of safety; an essential

technical and economic feasibility project, the final project and the executive one, which can be mutually combined also with the project's implementation.

[7]Eurovelo is a large European network of more than 70,000 km of cycling tourism paths. In Italy, a considerable part of Eurovelo path no. 8 corresponded to cycling route VENTO, which was, in turn, included in Bicitalia path no. 2, the network of itineraries and cycling tourism routes defined by the Italian Federation of Bicycle Friends. Project VENTO has partly reviewed the original route of Bicitalia 2, confirming its technical feasibility.

project-related decision made to work closely on the ambitious concept of cycling and pedestrian path (dedicated path), rather than on the ambiguous one of itinerary (shared path);

- attractive features: since the success of a main cycling tourism route is related to the images inspired, the landscape elements it connects and the technical and qualitative characteristics of the infrastructure itself; the design of the route had to create a network with the largest number of natural sites, of history, of culture, of farms, of medium-size cities, of hospitality and refreshment facilities, while the infrastructure had to be designed carefully (considering simple and functional architectural details) and, finally, it had to be long to allow a journey of several days for all users (with a safe and dedicated path);
- modal flexibility: it is preferable for a cycling path to be connected to the railway network (at a distance of not more than 5 km), an aspect which allows to extend the cycling catchment area, making segments usable by people of all ages and skills; the same applies for sailing on the river,[8] which makes the journey more attractive from a tourist perspective;
- simplicity: understood as the repetition of the same project solutions along the entire infrastructure by arranging an abacus of typological solutions, a crucial factor to construct an identity image of cycling and to limit implementation costs;
- affordability: to minimize implementation costs, the decision was made to develop the VENTO route alternatively on the left and right banks of the river Po. The objective was also pursued by using the main banks of the river, harnessing existing cycle paths, and limiting crossings (bridges) on the Po and on its tributaries.

These are the five outcome criteria of the research activity, corroborated by surveys along other European cycling routes. According to a study conducted by the European Parliament in 2012, the first two, safety and attractive features, are the key factors for the success of a cycling tourism infrastructure (Pileri et al. 2015). These criteria have allowed to carry out the feasibility study—in about 2 years, from January 2010 to May 2012—the main outcome of phase I, ideation, which closed with the public presentation of the study on 11 May 2012 at Politecnico di Milano. Invitations to the meeting were issued to contact persons of the two entities managing the river "River Po Basin Authority (AdBPo) and Interregional Agency of the River Po (hereinafter referred to as AIPo)", the regions crossed by the cycling route "Lombardy, Piedmont, Emilia Romagna, Veneto", Milan Municipality (because it was concerned by the connection with the exhibition area of EXPO 2015) and, finally, representatives of two associations "Touring Club Italiano, Italian Federation of Bicycle Friends" and other representatives of both institutional and non-institutional entities. Among the actors participating in various capacities in the project, some had already been involved in various ways during the study's processing phase.

[8] Sailing along the river Po can be experienced for more than half of its route: from Pavia to the estuary. The service is offered by few private operators who, at times, do not guarantee the transport of accompanying bicycles. River sailing is a very interesting theme, and project VENTO is studying it to establish a link with the cycling infrastructure.

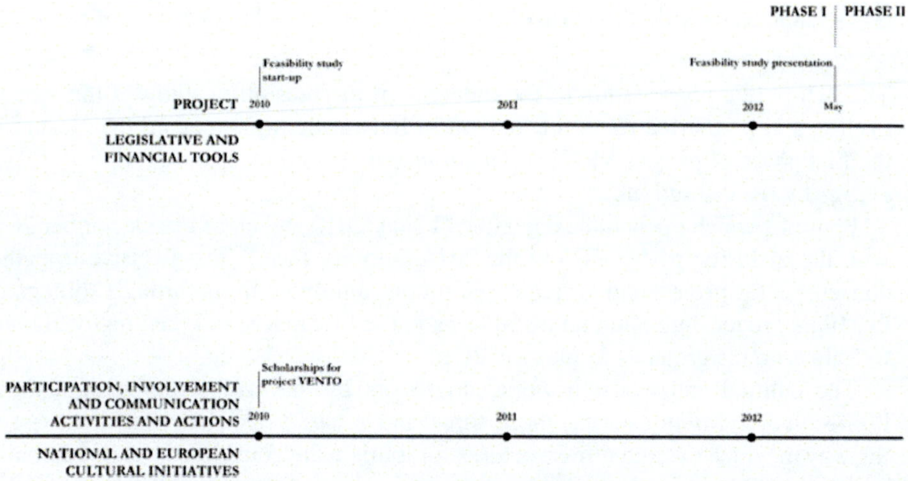

Fig. 4 Project timeline: phase I. The scheme represents the project timeline shared in three themes: legislative and financial tools; participation involvements and communication activities and actions; national and european initiatives from 2010 to 2020. *Source* the authors

In this period of time, the project was shaped and endowed with solid bases and credibility, both academic and public. At this time no actor was directly involved in planning. Focus was on indirect involvement with moments of academic discussion with other research teams (in Italy and abroad) that in various capacities managed the same themes, with interviews and targeted technical meetings that were essential for the project's actual feasibility, such as, for example, a discussion with AIPo on the use of cycle paths at the ridge of river banks. A decision inspired by the idea of maintaining the autonomy of the project's proposal within the disorganized scene of cycling tourism at the time (Fig. 4). The scheme represents the project timeline shared in three themes: legislative and financial tools; participation involvements and communication activities and actions; national and european initiatives from 2010 to 2020. This approach was at times hard to defend due to friction with certain actors who were poorly inclined to change, but which today we can deem winning thanks to the results achieved.

An extensive period during which the continuity of research work was ensured by a scholarship funded by the Lombardy Regional Government[9] and co-funded by the DAStU and by a private company (BLM Group).

[9]In December 2009 a scholarship provided by the Lombardy Regional Government was granted for research proposals related to the international exhibition EXPO 2015 held in Milan. Project VENTO, with the proposal of a cycle path to reach the EXPO area, suggested conveying visitors to the external section of the exhibition area to create an 'off-site EXPO' that would develop through territories with a powerful calling for farming: over 10,000 farms are crossed by the cycling route.

3 Phase II—Taking Charge

The first public presentation of the outcome of the feasibility study of the longest cycling and pedestrian route for tourism in Italy at the time marked the closure of the first phase of project VENTO, *ideation*, and opened the subsequent one, *taking charge* by the institutions.

Phase II, which opened the day after 11 May 2012 and closed in December 2015 with the inclusion of VENTO to the 2016 Stability Law,[10] was decisive to gather consent for the project and to create a common committed front, through Politecnico di Milano, requesting national institutions in the first place, and then regional ones, to follow-up the project's implementation.

The cultural and research work carried out by the multidisciplinary team of Politecnico di Milano during those three and a half years determined a twist in the way in which cycling infrastructures for tourism are considered today, compared to the past. The scientific research carried out achieved three main innovations:

1. today it is deemed that cycling tourism paths can be a trigger for occupational and economic development of fragile territories, which are excluded from traditional tourism but are rich in potential, if considered part of a broader linear/network system;
2. cycling tourism routes are, to all effects and purposes, infrastructures with precise performance requirements; itineraries, though duly marked, do not suffice to activate tourism flows capable of bringing important economic opportunities to fragile territories;
3. planning and creation of cycling tourism routes require a leap of scale, that is, local frameworks cannot be assigned the implementation of such projects. We need to move beyond local fragmentation, entrusting to a third party the design, implementation and management of these projects.

The research team of Politecnico di Milano, through intensive research work focused primarily on identifying and understanding successful foreign case studies, defined the issues to be brought to the attention of the public decision-maker in various scales. With significant culture-oriented work, the team built the consensus base that favours the project today.

Three tools are basically used in this phase:

- an adhesion campaign inviting residents, institutions and associations working in various scales (local, supra-local, national) to join the project; an e-mail for citizens, a letter on letterhead for associations and institutions, stating (without any subsequent economic commitment) their interest in the cycling route, and encouraging Regional Administrations, River Entities and the Government to initiate project implementation;
- signing of an Agreement Protocol, launched 1 year after the first public presentation of VENTO, to formalize the interest of local institutions in the project, and

[10]National financial predictive scheduling law.

their willingness to entrust it to a third party that would jointly manage the project implementation process;

- a travelling workshop, VENTO Bici Tour (held on a yearly basis since 2013), to:
    - o disseminate and propose, in territories crossed by the infrastructure, the essential cultural approach that underpins collecting consent for the project;
    - o present the local requests collected with the adhesion campaign and with the Agreement Protocol to the central institutions.

The adhesion campaign for the project and the signing of the Agreement Protocol have allowed to construct a wide common base around the project, which was crucial to approach and involve central institutions.

The adhesion campaign to VENTO, launched in June 2012, shortly led 110 leading local and national associations, 228 institutions of which 4 Regional Administrations, 11 Provincial Administrations, 185 Municipalities, 11 Parks, 18 other entities and little less than 5,000 residents to join the project. Instead, the Agreement Protocol launched in May 2013 with the contextual signature of the Milan Municipality, the City of Turin and of Venice, the Po River Basin Authority and Politecnico di Milano, was later signed by 167 institutions, of which 7 Provincial Administrations, 151 Municipalities (half of which directly crossed by the infrastructure and others, near and far from the route), 3 Regional parks and 6 other entities (Fig. 5).

The growing attention given to the project by local institutions, and their direct involvement in VENTO Bici Tour, the project's largest and most recurrent communication event, attracted the attention of the Ministry of Cultural Heritage and Activities and of Tourism, in the person of Minister Bray first (2013) and of Franceschini later (2014). During the Tour, they expressed their interest in the project and the intention to implement concrete actions to enter the subsequent planning phases of the infrastructure. However, we must wait for the third edition of VENTO Bici Tour (2015) for the project to be actually accepted by central institutions and, particularly, by the Ministry of Infrastructures and Transport. The former Minister, Hon. Delrio, participated in the institutional meeting organized in Cremona, announcing that "VENTO will be done!". This was followed by the inclusion of VENTO to the 2016 Stability

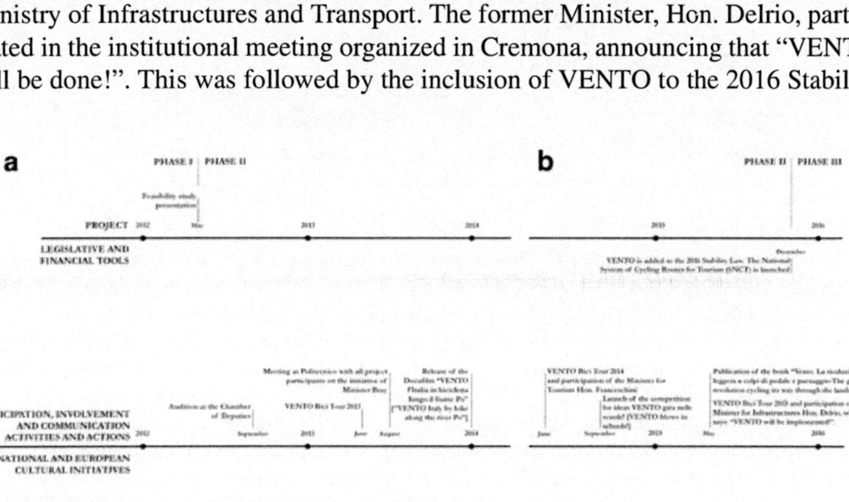

Fig. 5 Project timeline: phase II a–b. *Source* the authors

Law (Law no. 208 of 28/12/15, art. 1 section 640) as one of the four priority cycle paths, along with the Sole Cycle Route, the Apulian Aqueduct Cycle Route and the Great Bicycle Ring Road in Rome, which will become part of the National System of Cycling Routes for Tourism (SNCT) established by MIT and MIBACT (SNCT was increased to ten cycle routes by the subsequent 2017 Stability Law) (Fig. 6). To continue these projects, the Law has allocated 94 million euros for the next 3 years.

The 2016 Stability Law (approved in December 2015), which marks the first allocation of public resources for project VENTO, closed the taking charge phase of the project on the part of institutions. During this timeframe (May 2012—December 2015), the research team of Politecnico continued its work with the economic support of Bank Foundations. Fondazione Cariplo contributed to the project to continue research activities on complementary themes to the infrastructural project developed previously with the scholarship provided by the Lombardy Regional Government, and activated by the joint funds of Politecnico di Milano and of a private company.

Fig. 6 The National System of Cycling Routes for Tourism (SNCT). The first four cycling routes entered in the 2016 System:—VENTO Cycle Route, 680 km, from Venice to Turin;—Sole Cycle Route, 300 km, from Verona to Florence;—Apulian Aqueduct Cycle Route, 500 km, from Caposele (AV) to Santa Maria di Leuca (LE);—GRAB, Great Bicycle Ring Road, 44 km, in Rome. The 6 cycle routes entered in the System in 2017:—Garda Cycle Route, 140 km, along Lake Garda;—Magna Grecia Cycle Route, 1000 km, from Lagonegro (PZ) to Pachino (SR);—Sardinian Cycle Route, 1230 km, from S. Teresa di Gallura (OT) to Sassari;—Adriatic Cycle Route, 820 km, from Lignano Sabbiadoro (UD) to Gargano;—Trieste-Lignano Sabbiadoro-Venice Cycle Route, 150 km, from Venice to Trieste;—Tyrrhenian Cycle Route, 870 km, from the French border to Rome. *Source* Ministry of Infrastructures and Transport, and Ministry of Cultural Heritage and Activities and of Tourism

Compagnia di San Paolo, Fondazione Cassa di Risparmio di Padova e Rovigo and Fondazione Cassa di Risparmio di Bolzano, instead, contributed to the development of the main communication activities proposed by VENTO: VENTO Bici Tour.

In November 2014 Politecnico di Milano filed the VENTO trade mark (registration of the trade mark at the Office for Harmonization in the Internal Market—OHIM) to consolidate VENTO's identity and to preserve the founding principles of the territorial project proposed, besides guaranteeing the proposing subject technical-scientific supervision during the subsequent planning process.

4 Phase III—Implementation

The creation of the SNCT and the allocation of public resources for the subsequent planning and implementation processes by the Government changes project VENTO's implementation scenario, making it a project of nationwide importance. At the same time, as a result of the Government's particular awareness (particularly of the Ministries of Infrastructures and Tourism) of themes of sustainable mobility, between 2016 and 2018 Italy was particularly interested in creating the Atlas of Italian Hiking Trails (2017), designating 2016 as the "year of hiking trails" (and, later, 2017 as the year of "Italian small towns", 2018 as the year of "food", 2019 for "slow tourism" and 2020 for the "tourist train"). Three guidelines were also defined with strong slow mobility features: "Link up Italy. Strategies for Transport and Logistic Infrastructures" (2016), "Strategic Development Plan for Tourism 2017–2022" (2016), "Travelling in Italy. Extraordinary Plan for Tourist Mobility 2017–2022" (2017). In parallel, the State Property Agency launched the call for tenders "Hiking trails and paths" to encourage redevelopment of decommissioned real property along the main routes of the Atlas and SNCT cycling paths, including VENTO (analysed in detail in the part 4 of the book, Chap. 13).

In this rich context of changes, Politecnico di Milano decided to transfer, free of charge, the feasibility study of project VENTO to the Government on condition that the subsequent planning phase, the technical and economic feasibility project[6] (hereinafter referred to as PFTE), should be carried out jointly with a single call for tenders for the entire 700 km. This decision of the Politecnico was deemed essential for the efficient continuation of the project. The Government understood and pursued it, requesting the Regional Administrations concerned (Piedmont, Lombardy, Emilia Romagna and Veneto) to implement it. Moreover, considering the Politecnico's commitment to the project's ideation and to initiating local involvement, the University was also assigned the role of Technical-Scientific Secretariat at the Technical Round Table: the coordinating subject of the subsequent planning and implementation phases of the cycling route, comprising the four regions involved, Politecnico di Milano and the Responsible for the Project (ILSpA, inhouse company of the Lombardy Regional Government).

With the new role of Technical-Scientific Secretariat, the Politecnico's work continued in an entirely different setting. Not as sole/main actor of project VENTO

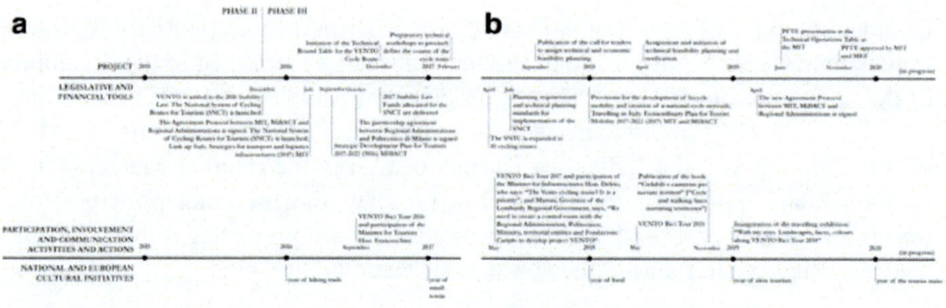

Fig. 7 Project timeline: phase III a–b. *Source* the authors

anymore, but as part of a team of actors made up of Regional Administrations appointed to implement the project and of the Responsible for the Project (hereinafter referred to as RUP) acting as Technical-Administrative Supervisor of the project's subsequent phase (Fig. 7).

On the one hand, regional institutions and the RUP were expert in programming–planning–designing road infrastructures, but not in cycling tourism infrastructures. On the other hand, we have an original territorial project in the Italian scene, paving the way for a new cultural approach to the theme of cycling tourism. In the middle we find the Politecnico facing a new challenge: to guarantee that the project's philosophy would be maintained with constant stories and argumentations addressing the many actors involved during the PFTE.

The main innovation in this phase was to successfully fill the cultural and technical gap in planning cycling tourism, with the work performed, within its competencies, by the Technical-Scientific Secretariat at the Technical Round Table starting from 13 December 2016, date of the first meeting. This important role has allowed to avoid project fragmentation and to maintain a broad and transparent common vision of the same with the support guaranteed throughout the planning process, particularly:

- support in organizing four moments to share the project's route (called workshops): the only instance in Italy, compared to the other cycle routes of the SNCT (February 2017);
- support to the RUP in organizing the international call for tenders[11] of the PFTE (September 2017);
- support for the designers[12] during the processing phase of the project (from April 2017 date of assignment of the PFTE, to November 2019) (Fig. 8);
- support during meetings organized for the preliminary Service Conference (April 2019), second moment of project sharing (after the workshops in February 2017); an action not requested by Italian law for this planning phase, but required by the

[11] It was a multi-lot competition to assign the PFTE of the cycle route VENTO and to verify and validate the same for a total amount of ca. 1.8 million euro (1,793,093.14 €).

[12] The PFTE was assigned to Temporary Project Grouping coordinated by Cooprogetti Società Cooperativa.

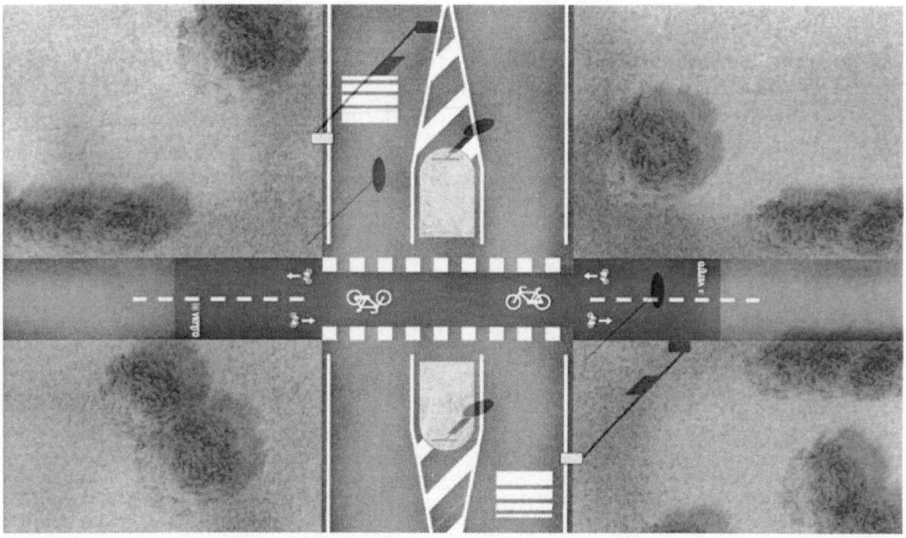

Fig. 8 Excerpt of some architectural and technical solutions designed in the PFTE. *Source* PFTE designers, 2019

Technical Round Table to address all concerns and to ensure the project was fully agreed with;

- support during meetings at the Ministry of Infrastructures to motivate the project-related decisions necessary for approval[13] by the Technical Operations Table made up of representatives of various Ministries and leading national associations (June, July and October 2019).

The implementation phase (in progress) is then extremely complex and rich in events that can be grouped into three temporal-theme lines. The first concerns planning the infrastructure, whose main activity is that of Technical-Scientific Secretariat as stated. The second is communication and participation mainly characterized by the annual editions of VENTO Bici Tour during which participating Government representatives confirmed the intention to continue project implementation. Particularly in the 2017 edition, Hon. Delrio (former Minister of Infrastructures) said, "The cycle route VENTO is a priority!". The third related to issuing technical and financial legislative tools to legitimize and ensure project feasibility, such as the Agreement Protocol of July 2016 between the Ministry of Infrastructures, Ministry of Cultural Heritage and Activities and of Tourism, and the four Regional Administrations crossed by VENTO; the 2017 Stability Law (Law no. 232 of 11/12/16, art.

[13]In November 2019 the Ministry of Infrastructures consented to the PFTE and to the four priority functional lots, one for each region, that will be subjected to the planning process (final and executive starting from January 2020), and will be finally created: from Chivasso to Turin for the Piedmont lot, from San Rocco al Porto to Stagno Lombardo for the Lombardy one, from Piacenza to Fossadello and the urban segment of Ferrara for the Emilia Romagna lot and from Adria to Loreo for the Veneto lot.

1 section 640), which integrates funds of the previous Law of 2016 for the SNCT; Law no. 2 of 2018, which lays down provisions for the development of bicycle mobility and the creation of the national cycling network and, finally, the Ministerial Decree Protocol 375 of 20 July 2017, which defines, in Enclosure A, the 'planning requirements and technical planning standards for implementation of the SNCT'.

During this phase the continuity of research work was guaranteed by the three funding trends: activity as Technical-Scientific Secretariat with a contribution of Regional Administrations; other research and communication activities with the initiative of VENTO Bici Tour through registrations of participants and funds from sponsors; while the planning and implementation of the cycle route VENTO is guaranteed by resources allocated by the 2016 and 2017 Stability Law amounting to 16,622,512.41 €.[14]

5 A Communication Project For a Territorial Project

Transversally to all the phases discussed—through which the research process and method have unfolded—it was essential to create a strong communication structure made up of miscellaneous languages, soft but at the same time rigorous to support and constantly accompany a complex territorial project such as VENTO, featuring cultural, technical and social density and diversity.

Communication is the theme capable of keeping together and of accompanying the various phases and dimensions of the project method implemented, with the objective of disseminating, in a complete, legible and easy to understand manner for everybody (residents, scholars, public administrators, etc.), its technical evolution and project philosophy. The communication project is not second to the infrastructural design project but is intrinsic to it. A project within a project that has built courageous images, which are, at times, original, based on miscellaneous languages and representations: facts, data, experiences, success stories, testimonies. Words, images and communication messages in a common playing field, always listening to the voice of the territories to build a common discussion that is choral and understandable to everybody.

Cultural communication developed by narrating VENTO through a dialogue with residents and institutions, local and national policies, thus providing a supplement of knowledge to the project phases and to the entire process. A communication that has become the tool of scientific dissemination, accompanying research for the creation of a recognizable and replicable model. A communication process that had to keep together the local and global framework with the original proposal of cycling, constantly working to fill any cultural gap and to provide learning tools for those who want to understand the reasons and the philosophy of the project.

Languages and communication methods have been many from 2010 to date, ranging from the exploratory phase to surveys, from public meetings in squares

---

[14] Interministerial Decree 29/11/2018 Plan for the division of resources.

Fig. 9 A social gathering of participants in VENTO Bici Tour 2019 in the municipality of Fontanetto Po: a good practice for project sharing. *Source* the authors, 2019

to photographic exhibitions. VENTO has thus addressed a heterogeneous public of subjects and successfully kept in step with the digital advent and its capacity to speak to many, narrating a piece of public history, even through social networks. In fact, from the onset of the project, the research team's intention was to attempt to explain, outside the academic setting, a project born in the university. The work entailed several multi-level communication actions, always underpinned by objective truths. Actions that were the glue to establish a dialogue between memory, tradition and current time; infrastructural design, places and social bonds (Fig. 9).

After completing the 2012 feasibility study, the research team visited municipalities and entities potentially interested in the project to explain its technical details and reasons for its implementation. A communicative and participative project presented in the field by the university to bridge the gap between academic theory and reality in order to broaden and consolidate the project sharing base.

The strongest and most innovative communications tool proposed is VENTO Bici Tour: a cultural bicycle tour through places touched by the project of the future cycling route, meeting institutions and citizens. By stepping into the field, we were able to analyse the details and study the territory, places and their stories, thus constructing a digital archive of pictures, voices and faces from the territory. Photographs and video footage transformed into audiovisual products created by professionals with perspectives changing over the years, disseminated over social channels and websites, projected into the municipalities at fairs, feasts and cultural events of all types (Fig. 10).

Fig. 10 One of the communication languages used for project dissemination: one frame from the video of VENTO Bici Tour. *Source* Marco Rimondi 2019

The physical presence of the research team in places crossed by VENTO was crucial and successful also as a result of a narration through the Web and social media, which today represent the project's most direct sounding board. After completing the feasibility study, a website was created, first www.progetto.vento.polimi.it, today www.cicloviavento.eu, followed by pages on social media: Facebook, Instagram and Twitter (@VENTOpolimi). And in 2017, with the advent of a VENTO Bici Tour open to all even in terms of bike riding (first it was only at events), a dedicated website—www.ventobicitour.it—was also created. Digital containers to which materials with different visual and narrative languages could be uploaded with the intention of simply and truthfully communicating knowledge to make the community feel part of this soft revolution (Fig. 11).

Moreover, projects and artistic practices of narration about the territory have been organized in these years, involving residents and local communities in a sort of osmosis between the territory, residents and cyclists.

#coloriamo vento[15] [let's colour vento] was launched in 2018, a project for public involvement in which local, social and cultural entities between the Municipality of Boretto and Guastalla committed to colour a segment of the cycling route VENTO between the two municipalities. Children from schools and disabled children from project K-lab of Reggio Emilia Città Senza Barriere were involved, with the cooperation of street artist Davide Bart Salvemini, using pavement graphics to create traffic signs dedicated to the project and painting part of the wall along the route. An excellent artistic and cultural public initiative to redesign the narration of a territory and to strengthen the feeling of community (Fig. 12).

[15] https://www.ventobicitour.it/coloriamovento/.

Fig. 11 VENTO website, www.cicloviavento.it. *Source* the authors, 2020

Fig. 12 Children and residents of the municipalities of Boretto and Guastalla during the participated activity of live painting #coloriamo VENTO along the course of the future cycling route in the Boretto municipality. *Source* author's name, 2018

"*Through my eyes. Landscapes, faces and colours along VENTO Bici Tour 2018*[16]", a travelling photographic exhibition (in progress) hosted in both formal and informal venues, born from a photographic contest organized during the Tour, which sees as protagonists the photographic views of participants transformed into seven self-bearing photographic wheels. Created with the artistic co-direction of the Museum of Photography of Cinisello Balsamo in partnership with the companies Campagnolo and Bianchi, and with the support of Fondazione Cariplo. The project continued in 2019 with "*Through my eyes. At pencil tip*[17]", a collection of postcards created by participants in the Tour to represent their vision of VENTO, experimenting with other expressive languages (drawing, writing) created in partnership with the company F.I.L.A. e Giotto—Colore Ufficiale (Fig. 13).

All these forms of communication through different narrative languages and a strong social involvement are the added value to understand and discover a project centred on the territory and on beauty, and to ensure its success. It is a way of communicating research and its contents with guiding and cultural actions for pedagogic and educational purposes.

An educational scientific communication, whose languages, precisely words, images and maps, is the tools that make research focused on converging the local

[16] https://www.progetto.vento.polimi.it/conimieiocchi.html.

[17] https://www.ventobicitour.it/con-i-miei-occhi-vento-in-punta-di-matita/.

Fig. 13 One of the laps of the photographic exhibition "Through my eyes. Faces, colours, persons along VENTO Bici Tour 2018" hosted at the Teatro Sociale of Luzzara. *Source* Alessandro Giacomel, 2019

and the global legible by using a variety of tools: scientific and informative articles, radio and Web interviews; books[18] and publications; websites and social channels; digital and hard copy graphic products accompanied by images (pictures and videos), numbers, diagrams and brochures dedicated to the tour and to the project, merchandising and slogans (Fig. 14).

The communications project played a key role in the accreditation process of the infrastructural project with the institutions at various levels and allowed project VENTO to reach its current level of development. Today it continues to play a key role in the process. Indeed, it is important for the university to maintain its role as spokesperson of a choral message from territories to central institutions.

6 A New Approach to a Territorial Project

With VENTO, the research team of Politecnico di Milano proposed, tested and refined a methodology to define territorial projects capable of converging multiple dimensions to be considered when pursuing development in complex contexts. The

[18]Pileri et al. (2015) VENTO. La rivoluzione leggera a colpi di pedale e paesaggio. Corraini editore. Pileri et al. (2018) Ciclabili e cammini per narrare territori. Arte design e bellezza dilatano il progetto di infrastrutture leggere. Ediciclo editore.

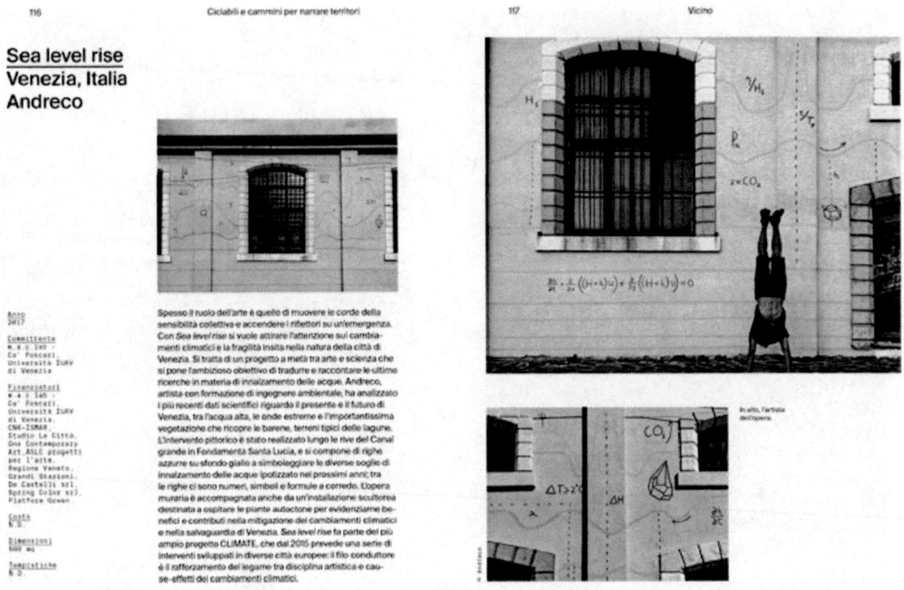
116 Ciclabili e cammini per narrare territori

117 Vicino

Sea level rise
Venezia, Italia
Andreco

Anno
2017

Committente
M.A.C. lab - Ca' Foscari, Università Iuav di Venezia

Finanziatori
M.A.C. lab - Ca' Foscari, Università Iuav di Venezia, CNR-ISMAR, Studio La Città, One Contemporary Art, ASLC progetti per l'arte, Regione Veneto, Grandi Stazioni, De Castelli srl, Spring Color srl, Platform Green

Costo
N.D.

Dimensioni
600 mq

Tempistiche
N.D.

Spesso il ruolo dell'arte è quello di muovere le corde della sensibilità collettiva e accendere i riflettori su un'emergenza. Con *Sea level rise* si vuole attirare l'attenzione sui cambiamenti climatici e la fragilità insita nella natura della città di Venezia. Si tratta di un progetto a metà tra arte e scienza che si pone l'ambizioso obiettivo di tradurre e raccontare le ultime ricerche in materia di innalzamento delle acque. Andreco, artista con formazione di ingegnere ambientale, ha analizzato i più recenti dati scientifici riguardo il presente e il futuro di Venezia, tra l'acqua alta, le onde estreme e l'importantissima vegetazione che ricopre le barene, terreni tipici delle lagune. L'intervento pittorico è stato realizzato lungo le rive del Canal grande in Fondamenta Santa Lucia, e si compone di righe azzurre su sfondo giallo a simboleggiare le diverse soglie di innalzamento delle acque ipotizzato nei prossimi anni; tra le righe ci sono numeri, simboli e formule a corredo. L'opera muraria è accompagnata anche da un'installazione scultorea destinata a ospitare le piante autoctone per evidenziarne benefici e contributi nella mitigazione dei cambiamenti climatici e nella salvaguardia di Venezia. *Sea level rise* fa parte del più ampio progetto CLIMATE, che dal 2015 prevede una serie di interventi sviluppati in diverse città europee: il filo conduttore è il rafforzamento del legame tra disciplina artistica e cause-effetti dei cambiamenti climatici.

Fig. 14 Narration of the territory through some pages from the book "Ciclabili e cammini per narrare territori" ["Cycle and walking lines narrating territories"], authored by the research team. *Source* the author

infrastructural dimension is one of the aspects of the project that cannot exist without parallel project-based focus on the historical, landscape and environmental context in which the infrastructure is placed, and without cultural and training actions addressed to those who live in the territories concerned by the project. Starting from tomorrow, they will be the key actors of the development process that the territorial project wants to trigger.

The territorial project thus conceived constantly unfolds on two scales: local and territorial. It considers local resources but reassembles them in a territorial plan capable of preserving the local identity and of enhancing its role. It discusses issues with local institutions but asks them to perform a leap of scale, abandoning localism and embracing a new perspective to reinterpret itself as part of a larger system. It overcomes the administrative limits in which the territorial government has been bridled, reconfiguring new project geographies capable of creating innovative inter-institutional alliances. It places the project back in the hands of institutions, which it deems play a key role in the process, namely, regional and central institutions, in order to guarantee their coordinated implementation and to preserve their individuality, both from a technical standpoint and in terms of identity.

The aspects discussed constitute the backbone and the key ingredients that characterize a territorial project. They must be considered if one wishes to replicate the model in other contexts or to extend it to other project categories.

VENTO has triggered a cultural change that has involved multiple subjects: institutions at various scales and roles (technical/political), subjects with wide-ranging interests and citizens.

In institutional terms, VENTO has requested local institutions to take a step backwards as regards decisions on a territorial scale, and to rely on a new "subject" that has intervened to guide local requests, create a system with them and bring them to the attention of regional and central institutions. VENTO asked the regional institutions to maintain a unitary vision of the project, avoiding fragmentation and attempting to reposition its competencies by rising from a subject that administers funds to a subject that plans territorial development in a rationale of vision, strategy and action. VENTO has asked central institutions to allocate funds for the development of territorial projects capable of creating development in fragile territories, contributing to the birth of the SNCT and to the introduction of new regulatory devices on the subject of sustainable mobility.

As regards subjects with widespread interests, miscellaneous associations and professionals in various fields of technical expertise, the cultural change triggered by VENTO once again concerns the scale in which each subject interprets his work and the competencies it can contribute. Even local associations must bring a territorial vision and their work is part of a broader plan. Technical know-how requires ongoing training, specificity, and the ability to cross the various scales of the project. Designing long-distance cycling routes for tourism is quite another matter to planning cycle paths in an urban context. Hence, it requires the development of dedicated professional expertise.

Finally, with the cultural work carried out, VENTO has informed even citizens about the issues, explaining that cycling tourism is a practice for everybody, a practical and viable option to relaunch both employment and economies in fragile areas. It has summoned each one to their role as active residents capable of addressing a clear question to the institutions that govern the territory.

VENTO has perfected a new method for participated planning in the framework of territorial projects. It is not centred on project sharing at a micro-scale but rather on sharing high ideals that drive the project for a great work through another major work.

Project VENTO has come a long way in these 10 years, but there is still a lot to be done to implement the infrastructure. The research team of Politecnico di Milano will do its utmost to remain within the planning process of the infrastructure to guarantee compliance with project criteria that inspired the infrastructural project, and to accompany its implementation as a territorial project and not only as an infrastructural one. Hence the need to continue the research activity to shape a tourism-oriented project that is complementary to the infrastructural one, and to perfect a unitary management model of the infrastructure that precedes its implementation to guarantee its future function.

References

Alberti F, Munarin S, Pileri P, Zazzi M (2019) Oltre l'automobile. Forme innovative di mobilità per la rigenerazione urbana e territoriale. Donzelli, pp 219–226

Dente B (2011) Le decisioni di policy. Come si prendono, come si studiano. Il Mulino editore

Pileri P, Giacomel A, Giudici D (2015) VENTO. La rivoluzione leggera a colpi di pedale e paesaggio. Corraini editore

Pileri P, Giacomel A, Giudici D (2017) VENTO. Pedaliamo sulla terra leggeri: VENTO come progetto di territorio. Urbanistica Informazioni 272:17–19

Pileri P, Giacomel A, Giudici D, Munno C, Moscareli R, Bianchi F (2018) Ciclabili e cammini per narrare territori. Arte design e bellezza dilatano il progetto di infrastrutture leggere. Ediciclo editore

Narration of Cultural Heritage as *Antifragile* Tool

Catherine Dezio

Abstract In our contemporary, multicultural and globalizing society, what can help a fragile territory and community in the protection of its cultural heritage? While walking through a territory, you can realize how it offers itself as a story in continuous elaboration, able to give back ancient and authentic memories, but also to build new ones that can generate opportunities. In this sense, the ancient and spontaneous practice of narration can become an innovative *antifragile* tool: as an inclusive action for marginal individuals; as a new protection practice for small museums; as an opportunity that facilitates local development processes of depopulated territories, for tourists but also for communities themselves. After a discussion on the topic of narration and its *antifragile* abilities, we will describe the case of VENTO crossing Casale Monferrato, a small town in the Italian region of Piedmont. VENTO is a project for a cycle route that crosses northern Italy and hopes to mend the identity weave of depopulated territories, generating jobs through the narrative ability of the slow line.

1 Narrating Cultural Heritage

"Every stone represented the singular conglomerate of a will, of a memory, sometimes of a challenge. Each building stood on the plan of a dream" (Marguerite Yourcenar, *"Memoirs of Hadrian"*, 1951). This sentence by Marguerite Yourcenar tells of a significant and profound change in the concept of cultural heritage. From the documentation and protection of cultural heritage, today there is a passage towards heritage as a source of new knowledge. The cultural heritage is no longer intended as an asset in itself, to be preserved or enhanced, but as a source of multiple, integrated, relational information, full of emotion and nostalgia for eras lived or sometimes never lived. This dynamism also constitutes the possibility of identifying cultural heritage where we did not expect it before: outside museums, in unexpected, unknown, forgotten places. Between this new vision and a new user (on one hand

C. Dezio (✉)
Department of Architecture and Urban Studies, Polytechnic of Milan, Milan, Italy
e-mail: catherine.dezio@polimi.it

P. Pileri and R. Moscarelli (eds.), *Cycling & Walking for Regional Development*, Research for Development, https://doi.org/10.1007/978-3-030-44003-9_12

increasingly demanding, on the other increasingly ignorant), there is an innovative tool for returning complex content in a simple way: narration. Narration is the tool that accompanies this cognitive revolution, where the understanding is an experience. Since narration entered the museums, it is proving to be not only a flexible and congenial tool for mediation of cultural heritage, but a real resource for citizenship, critical literacy and the regeneration of identity belonging (Brunelli 2011; Bodo et al. 2016). It is about the quote of Yourcenar that tells the memory and the will behind a stone, recalling an image of a space, a story, an emotional load. Transporting this synergy between oral, visual and spatial narration in places means generating new forms of knowledge and communication. The didactic component mixes with the evocative one; the official story intertwines and declines in personal stories; the scientific language is combined with the informal one; the memory contained in museums has the same value as the memory presents in the street. Narration is an innovative but also ancient tool. Humans are creatures of communication: not only our species is a communicating species that uses words to narrate since ancient times, but since ever communication intertwines with emotions, building and refining our intelligence (Longo 2008). Stories reconstruct the world and ourselves in the world, so much that in some cultures the world does not exist before being narrated. It can be said that narration is an activity necessary for man to communicate knowledge to other men and to give sense to past events and objects. So necessary and so simple at the same time. To have a narration it is sufficient a narrator, a story, someone to tell it to. A narrator has information that he chooses to share *generously* with other individuals (Bernardelli 1999). It is a thread that unites. It is a thread that unites oneself, tracing those *"cultural intimacies"* (Herzfeld 2011; Zingari 2015), which today are needed to face globalizing models and are claimed as a human right to *"cultural existence"* (Tornatore 2011; Zagato 2014). It is a thread that unites experience of individuals with a wider community, generating sharing. It is a thread that unites generations of today with past and future generations, acting as a fundamental pedagogical tool for self-awareness (*"We are our past"*, said Henri Bergson). And it is also a thread that unites individual objects and stories, material and immaterial heritage, in their original temporal and spatial location. Through narration takes shape this heritage of memories, which defines a tradition and a cultural identity, individual and collective at the same time. Each narrator is invested with the difficult task of creating a complex narrative world, composed of characters, objects, events, giving them well-defined spatial and temporal coordinates. This is how that Yourcenar's stone acquires meaning, being placed back in its history, telling its origin and the reasons for its essence. Narration is basically a primitive and necessary act. It is necessary to give a sense to the world, a sense that exists before we perceive its presence.

2 An *Antifragile* Tool

Remembering that the notion of *heritage* is strongly connected with that of *territory* and of *community*, narration could be a versatile tool for a new protection of heritage in contemporary dynamic multicultural societies.

In order to always recognized a value for heritage, it must be able to renew itself, with a new aspect: a social aspect as well as cultural aspect (Salerno 2013). In this sense, narration has the potential to become an *antifragile* tool. Firstly, it can be *antifragile* for a new inclusive education. Some categories of users can benefit from the narration of cultural heritage: ordinary people, who have lost awareness of the value of history, and especially young people, who are the first with the task of passing on heritage to the future; migrants, usually excluded from the world of heritage; disabled, who often are unable to access to the variety of culture. These and other *fragile* actors may be the first beneficiaries of narration. *Knowing* is the tool to *include*; *learning* means being able to *remember* (Recalcati 2014); *memory* is the first step towards *protection.*

In parallel, narration can be *antifragile* also for territories.

Starting with the first heritage sites, museums, they are called to renew their role, today more than ever. We need active cultural centres, with variable and welcoming borders, which establish a constructive relationship with their visitors and which are able to become key points for territories.

We are talking about the big museums of metropolises, but above all we are talking about the small museums in marginal contexts. Small Italian museums are often unknown, widespread in depopulated territories, but they carry with them stories and identities.

For example, there is the Piccolo Museo del Diario (Small Museum of the Diary), inside the Archivio Diaristico Nazionale (National Diaristic Archive) of Pieve Santo Stefano, a small Tuscan village crossed by the itinerary Via di Francesco. There, Saverio Tutino (journalist and partisan during the Italian Resistance) collected memories and stories of Resistance through personal diaries. In an era where everything is digital, smart, fast, a place where you can touch, read, get lost in personal lives located in history is an extraordinary act, an other act of Resistance.

The 1914–1918 Museo della Guerra (War Museum), in Passo Fedaia in Canazei, meets hikers on the Marmolada glacier, a place of battle between Italian and Austro-Hungarian troops during the first conflict. Here Andrea Bernardin, an expert on the Great War, has collected 700 objects from the Marmolada mountain. In 2002, he chose to close the family restaurant and set up spaces for a permanent exhibition.

Among the Sienese hills there is the Museo dell'Antica Grancia (Museum of Ancient Grancia), originally built in 1200 as house for pilgrims of the Itinerary Via Francigena. Today is a museum of olive oil and of documentation of the "Grance" (from the ancient Latin "*granary*"), which tells the story of the ancient peasant life through objects and utensils of everyday life.

These are just some of the many small museums spread throughout Italy and the many forms of narration make them value. It is the narration of an object that finds its historical thread, or of an event that has changed the identity of a place. It is also the narration of the story of people who collect, preserve and who do not want to make a memory die. And finally, there is also the relationship between different narrative systems: the scale of object that finds its history in the scale of museum, located on a territory, that also asks to be told.

Donadieu (2013) tells the multiple virtues of narration on territorial scale. The narration can discover buildings that symbolized parts of community life, religious (churches, chapels, sanctuaries) or profane (fortified cities, castles, mills), and through them it can mark the agricultural or forest landscapes as signs of a collective memory (Donadieu 2013). The narration can also use literary, pictorial or photographic works that describe landscapes to discover unknown territories. Or the narration can give voice to the inhabitants, who highlight their memories, placing them back in the thread of history. It is at this level of popular expression that ordinary places are cultivated: fields and gardens rise again because someone's word has identified them, an existential word that sets in motion sufferings and pleasures, frustrations and desires, disappointments and joys.

As many narration as the dimensions of space and time. All of them have the faculty of being *antifragile* tools for marginalized individuals, small elderly societies, unknown heritage, disappeared memories, depopulated territories, as long as they are guided by the intentionality of a visionary project.

3 A Narrative Line. The Case of VENTO

There are territorial projects that work to narrate territories. Cycles routes and paths seem to be created to connect two opposite places, but actually they act as zippers. They are lines that unite: they unite the two edges of territory they cross and they unite the individual and punctual elements. The line of a cycle route is a light thread that *"works as a hidden supporting structure, an iron thread that can support the fragile tales that are deposited in the territory you cross, in the landscape you admire, in the scents you feel, in the faces you meet, in the flavours you can taste, in the stories that are kept behind a door, in the crafts that resist, in the colours of the walls that you touch, in the variety of cultivated fields, in the coolness of a forest, in the footsteps of who before have in turn told those places"* (Pileri et al. 2018). Line tourism has the potential to be *"pedagogical narration for those who pass by and ontological for those who inhabit"* (Pileri et al. 2018).

In 2010 a case study was born that is suitable for telling the narrative skills of the slow line. VENTO is a territorial project conceived and developed by a group of researchers from the Department of Architecture and Urban Studies of Politecnico di Milano (Italy), which is based on the power of the line: it is a cycle route that follows the Po river in northern Italy and connects Venice to Turin. In 2015, VENTO entered the Stability Law (Law n. 208 of 28/12/15, art. 1 paragraph 640), being recognized as one of the priority cycle paths of the National System of Tourist Cycle Routes (consisting of Ministry of Infrastructure and Ministry of Cultural Heritage), and it is arrived today to the feasibility project. 700 km of cycle path passes through villages and small municipalities in depopulation. Italy is a country of villages, where the 70% of Italian municipalities can be defined as *"small"* (i.e., with less than 5,000 inhabitants, according to the official definition of National Association of Italian Municipalities), and cover the 54% of the territory (IFEL 2011; Pileri and Granata 2014; Casa e Pileri 2017). Furthermore, according to the National Strategy of Inner Areas, more than 60% of the Italian territory is occupied by inner areas, defined as *"areas significantly distant from the centres offering essential services, but rich in important environmental and cultural resources and highly diversified by nature"* (SNAI 2013). The scarce offer of services (education, mobility and health care), accessibility and job places is the effect of a series of different dynamics that have determined the uneven development between city and countryside, mountains and plain, coast and hinterland (Marchetti et al. 2017; De Rossi 2018).

These territories ask to not be forgotten and look for new possibilities for rebirth. VENTO tries to work for them, setting the goal of *"sew up beauty of the territories crossed, reviving their vitality"* (Pileri et al. 2018). This is how VENTO hopes to narrate, educate, save. It aims to train people to be aware of contributing to triggering virtuous processes, recovering abandoned assets, activating new jobs and saving places from *"an unjust destiny of oblivion"* (Pileri et al. 2015).

Pazzagli spoke of beauty and associates it with the taste (*Belpaese* and *Buonpaese*, Pazzagli 2014). Around this concept, a new phase of development can be designed, which focuses on the quality of time, landscape, air, water, culture, food, life. All this is the philosophy of slowness. Nietzsche in "Human too Human" (1878) says: *"the noblest kind of beauty is that which does not transport us suddenly, which does not make stormy and intoxicating impressions (such a kind easily arouses disgust), but that which slowly filters into our minds"*. The adage philosophy is the main ingredient for the narrative power of VENTO, which uses slowness as a lens of knowledge. Here the narrative project looks for new forms with which to give voice to the places crossed, allowing them to tell their proper story. Local stories or popular wisdom, words, memories, phrases, proverbs, recipes, trades, can find space. They are Individual letters related to each other that a thread can tie together in a speech (Pileri et al. 2018). In this sense, the *antifragile* capacity of narration is fully realized, making the hidden identity patterns recognizable, revealing and enhancing the characters of beauty, and generating jobs through new local microeconomics.

Therefore, we talk about an action that has the power to be multifaceted and culturally and socially powerful. A sustainable, responsible, ethically oriented tourism (Nocifera et al. 2011; Pavione 2016), able to use narration to impart an education for caring and protection. This suggests the possibility of a tourism that reverses the concept of heritage, from something that is *consumed* to something that *contains us* (Montanari 2014), to be taken care of in the most emotional sense of the term. It will be a narrative tourism that is capable of educating and training, caring for and having care, repairing and regenerating, giving vision and changing horizons.

4 VENTO in the Casale Monferrato Area

We said that VENTO crosses 700 km of small towns, many stories, tangible and intangible heritage, that require attention. We are talking about a territory that asks to find a new identity, thanks to a research opportunity with the Province of Alessandria, in the Piedmont region.

We are in the VENTO section that falls within the Province of Alessandria, a predominantly agricultural area which includes 16 municipalities: the most populated municipality, that is Casale Monferrato (34,812 inhabitants), the second most populated, that is Valenza (18,634 inhabitants), and other 14 small municipalities around them, with less than 2,000 inhabitants and in the depopulation phase (12 out of 16 have a negative population variation rate; data source: ISTAT 2001–2013).

In particular, here we will consider the municipality of Casale, as the most reference point for this territory in terms of cultural identity and economic system, even if in constant depopulation. In 1981 Casale had 41,899 inhabitants; in 1997 there were 37,493 inhabitants; in 2019 there were 34,812 inhabitants (source: ISTAT). The municipality of Casale Monferrato has an extension of 86.21 km^2 and is located between the cities of Vercelli, Alessandria, Asti and Novara, in the industrial triangle of large cities of Turin–Genoa–Milan.

From a geographical point of view, Casale extends over a flat area located at the foot of the Monferrato hills, crossed by Po river and manned by "Po Park Vercellese Alessandrino". Recalling what Turri described as landscape icons, or *"elementary units of perception"* (Jodice and Turri 2001), the image of the local landscape is given by some characterizing elements: the most common crops, namely, rice fields, poplars and arable land; the local building typologies, first of all the *Cascinas* (farmhouses of northern Italy) and some mills still partially functional; native flora and fauna.

Even from an economic point of view, Casale is located at an intersection.

There is the historical economic activity of the place, that of rice cultivation, linked to the irrigated landscape of the rice fields. Then, there is the wine-growing activity of the hills, which has long been in decline due to the abandonment of marginal land. But the main economic activity of the place is the production of cement, which has also generated the most incisive identity image. For many years Casale has been known for the presence of the Eternit factory, now in disuse, which dispersed asbestos dust in the surrounding environment. From the 1950s to the present day, the inhalation of this material has caused more than 2,500 deaths and even today there is a high mortality rate attributed to Eternit among people born in the late 1980s, when the factory was dismantled. The phenomenon, in addition to the roofs of houses, sheds, churches, also affects the streets. All the streets of the municipalities near Casale were paved with this material, even if covered with a thin layer of asphalt. The long legal path is still ongoing with the "Eternit Bis" trial, which takes place in four separate forums, and the charge is of multiple aggravated manslaughter. Meanwhile, the peak of dead is expected from 2020 onwards, even if the decontamination is continuous.

Casale does not want to forget the drama, but at the same time it asks for a possibility of reborn.

To understand how narration can be salvific and can trace new and regenerative identity patterns, a survey was conducted that uses VENTO as an opportunity to discover new stories to tell. In this sense, a Food Heritage Map has been created, whose VENTO thread is the reference narrative system (Fig. 1, built starting from Novellini 2002 and Barbero 2002). The map collects the food heritage, both as local raw materials, determined by the climate and the conformation of soils (for example, due to the presence of the river: amphibians, molluscs and fish), both as recipes and products linked to the Piedmontese territory (*Krumiri* biscuits, *bagna cauda, agnolotti Casalesi,* etc.).

This map has food as protagonist but it also contains the many ways of preparing it (result of the countless hybridizations), the stories of those who prepare it and the landscapes that produced it. All of these factors become a single territorial intangible heritage.

Then there is a second Map (Fig. 2, built starting from Barbero 2002), which tells different fragments of the intangible heritage linked to the rural world. A world of stories, of folk dances and songs, of proverbs related to country life, of festivals and customs.

There is the Mondina figure, very common in northern Italy between the end of the nineteenth century and the first half of the twentieth century, now disappeared. She was a seasonal worker in the rice fields which, during the flooding of fields from late April to early June, they were staying for whole days with water up to their knees, barefoot and with a curved back to remove the weeds that grew in the rice

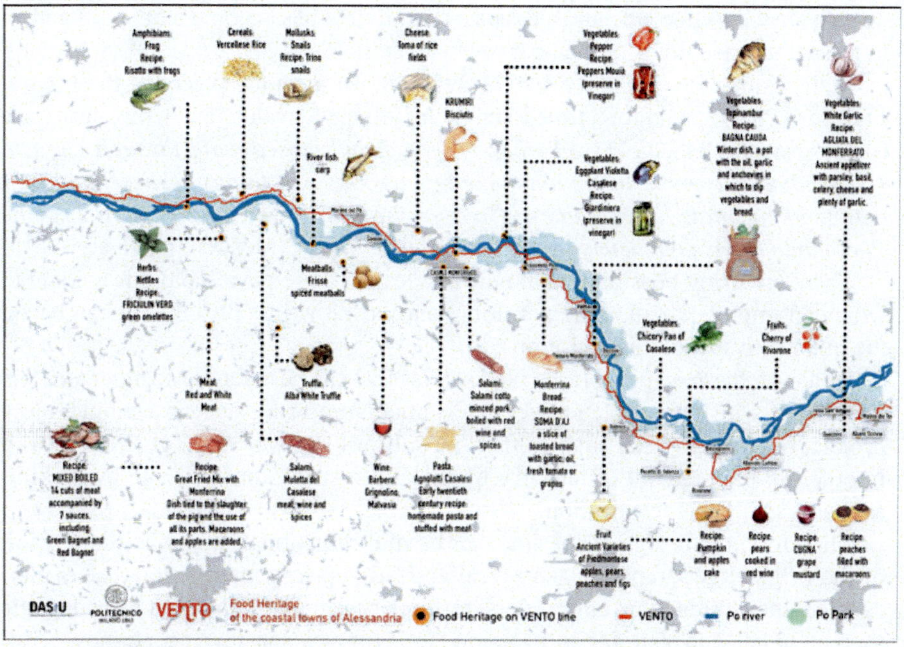

Fig. 1 Food heritage map. Elaboration of the author, 2019

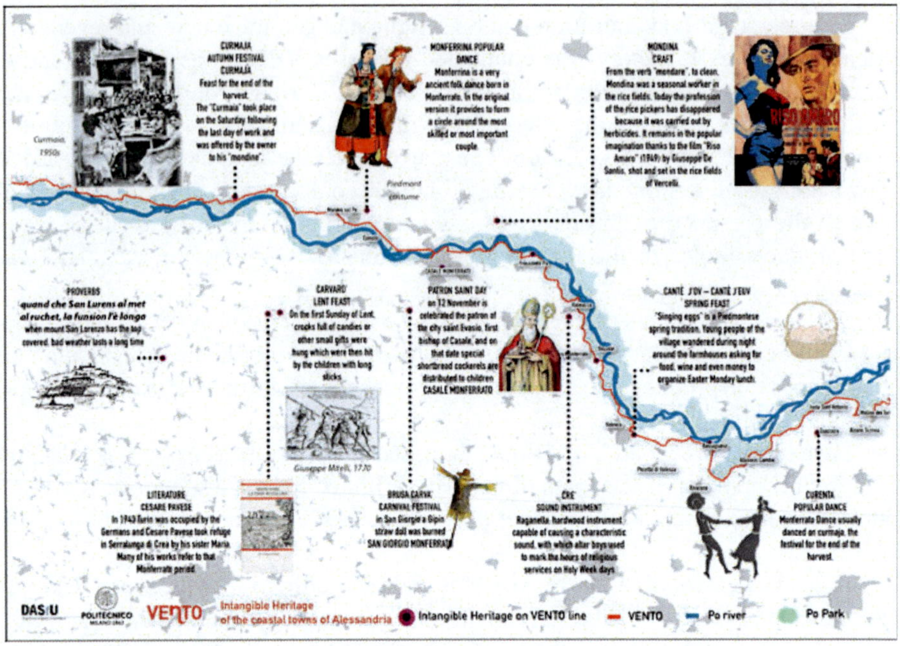

Fig. 2 Intangible heritage map. Elaboration of the author, 2019

fields. It was a tiring job, with bad working conditions. The 1949 film "Riso Amaro", directed by Giuseppe De Santis, draws a portrait that has become iconic, set and shot in the nearby rice fields of the near town Vercelli.

In the same map, along the VENTO thread, we meet the places of the literature of Cesare Pavese, who for a time lived in the village Serralunga di Crea, near Casale (in the period 1943–1945). *"Perhaps the sad and closed walk through Crea told you symbolically more than many people, passions and things of these months. Of course, the myth is a discovery of Crea, of the two winters and of Crea's summer. That mountain is completely impregnated by it"* (from "Mestiere di Vivere", 1952).

There are many possibilities of new images for these places and it is evident that what can improve them is the narration, and the slow line of VENTO can become the main activator of this narration.

However, these maps represent an exercise that can be extended to other territories and routes, with a view to a project that is not commercial, such as a simple tourism project, but pedagogical, such as free lines. Lines such as VENTO allow us to put in order all the material of territory by making it speak for itself and by making it available to a genuine narration.

Therefore, the design of the line can be multifunctional: it can help the reading of the original geographies; can reveal and tidy up forgotten heritage elements; it can become a thread that embroiders, restores broken tales and attracts pedagogical tourism to educate, care for and generate jobs for new sensitive economies.

5 Lesson Learned

Saramago says that *"the journey never ends. Only travelers end. And they too can extend in memory, remembrance, narration"* (from "Viagem a Portugal", 1981). That's why narration is important: to continue living and making people live through memory. The etymology of narration (from latin *gnoscere*) suggests the notion of knowing and making known. Narration produces knowledge and *"to know is an infinitive verb"* (Granata et al. 2013). And the magic of narration is endless, since it is capable of creating heritage of memories from single objects and providing, to those who have listened to them, the magnificent opportunity to keep them alive by becoming their new custodian.

Pasolini in 1975 said *"defend the mulberry pickets, the alder, in the name of the Gods, Greek or Chinese. Die for your love of vineyards. (…) Defend the fields between town and countryside, with their abandoned cobs. Defend the meadow between the town house and the canal. (…) defend, protect, pray!"* (Pasolini 1975). "Saluto e augurio" (1975), originally in Friulian dialect, is a testament addressed to a bourgeois boy, a student who still loves the ancient knowledge that his peers have rejected. Pasolini asks precisely to an interlocutor distant from his way of being to defend the past and all values that belong to the agricultural world that is being lost. *"Defend, protect, pray"* are actions of a project for a revolution of will (*defend*), of mind (*protect*) and of soul (*pray*). It is a project that addresses the furthest interlocutor, since those who do not know have the opportunity to preserve an almost disappeared heritage, handing it down.

Also Vito Teti reminds to the reader a relationship based on soul and on a sense of plural belonging: *"We are our body, our family, our ancestors, our history. We are the place where we were born and raised and the inhabited, known, lived places. The places dreamed and desired; but also places we fled from, or despised, loved and hated places. The concept of place cannot be simply understood in terms of space: (…) every place is mental and requires a symbolic organization that has to do with time, memory and oblivion"* (Teti 2018).

What is common to thoughts of Pasolini and Teti is that for both each of us can become the narrator of *"what remains"* (Teti 2017) and create *"minimalist utopias (..) even without heroes"* (Zoja 2013).

All this described up to now is a store of reflections, potential and hopes that asks to be realized by individuals, by community, by local administrations, beyond the media banality of typical products, festivals and folklore. We must work together for material and immaterial heritage, recognized and unrecognized, inside and outside museums, universities, libraries, archives, to reconstruct a new, powerful, intentional and *antifragile* narration. It will be a narration able to show that the knowledge that we thought dead, actually it is alive and is capable of including individuals, uniting the community, generating sensitive economies and is capable of reconstituting a past history knowing how to contain the future one.

References

Archivio Diaristico Nazionale. http://archiviodiari.org. Accessed 2 Jan 2020

Barbero A (2002) Camminare il Monferrato. Guida al trekking in Monferrato in 40 percorsi. Editrice il Monferrato, Villanova Monferrato

Bernardelli A (1999) La narrazione. Laterza

Bodo S, Mascheroni S, Panigada MG (2016) Un patrimonio di storie. Mimesis Edizioni

Brunelli M (2011) Heritage interpretation. Un nuovo approccio per l'educazione al patrimonio. EUM, Macerata

Casa M, Pileri P (2017) Il suolo sopra tutto. Cercasi "terreno comune": dialogo tra un sindaco e un urbanista, Altreconomia, Milano

De Rossi A (2018) Riabitare l'Italia. Le aree interne tra abbandoni e riconquiste. Donzelli Editore, Roma

Donadieu P (2013) L'immaginario dei territori agrourbani o la terra ritrovata. Scienze del territorio 1/2013. Firenze University Press

Granata A, Granata C, Granata E (2013) Sapere è un verbo all'infinito. Ed Il margine. Trento

Herzfeld M (2011) Cultural intimacy. Social poetics in the nation-state. Routledge

IFEL (2011) Atlante dei piccoli comuni. Fondazione IFEL, Roma

Jodice M, Turri E (2001) Gli iconemi: storia e memoria del paesaggio. Electa, Milano

Longo G (2008) Il senso e la narrazione. Springer

Marchetti M, Panunzi S, Pazzagli R (2017) Aree interne. Per una rinascita dei territori rurali e montani. Rubettino Editore, Soveria Mannelli

Montanari T (2014) Istruzioni per l'uso del future. Il patrimonio culturale e la democrazia che verrà. Minimum Fax

Museo dell'Antica Grancia. http://www.museisenesi.org/musei/museo-dellantica-grancia-e-dellolio.html. Accessed 2 Jan 2020

Museo della Guerra 1914-1918 di Passo Fedaia a Canazei, http://www.lagrandeguerra.net. Accessed 2 Jan 2020

Nietzsche F (1878) Umano troppo Umano

Nocifera E, de Salvo P, Calzati V (2011) Territori lenti e turismo di qualità. Prospettive innovative per lo sviluppo di un turismo sostenibile. Franco Angeli, Milano

Novellini S (2002) Dispensa del Po. I sapori del grande fiume. SlowFood

Pasolini P (1975) La nuova gioventù. Garzanti, Milano

Pazzagli R (2014) Il Buonpaese. Territorio e gusto nell'italia in declino, Felici Editore

Pavese C (1952) Il mestiere di vivere: diario 1935-1950

Pavione E (2016) Turismo sostenibile e valorizzazione del territorio. Sfide e opportunità di sviluppo del cicloturismo in Italia. Aracne Editrice, Ariccia (RM)

Pileri P, Giacomel A, Giudici D (2015) VENTO. La rivoluzione leggera a colpi di pedale e paesaggio. Corraini Edizioni

Pileri P, Giacomel A, Giudici D, Munno C, Moscarelli R, Bianchi F (2018) Ciclabili e cammini per narrare territori. Ediciclo Editore, Portogruaro (VE)

Pileri P, Granata E (2014) Piccoli comuni, grandi responsabilità. In Bonini G. e Visentin C. (A cura di), Paesaggi in trasformazione teorie e pratiche della ricerca a cinquant'anni dalla Storia del paesaggio agrario italiano di Emilio Sereni, Editrice Compositori, Bologna

Recalcati M (2014) L'ora di lezione. Per un'erotica dell'insegnamento. Super et opera viva

Salerno I (2013) "Narrare" il patrimonio culturale. Approcci partecipativi per la valorizzazione di musei e territori. Rivista di Scienze del turismo, 1-2/2013

Saramago J (1981) Viagem a Portugal

SNAI (2013) Strategia Nazionale per le Aree Interne. http://www.programmazioneeconomica.gov.it/2019/05/23/strategia-nazionale-delle-aree-interne/. Accessed 14 December 2019

Teti V (2017) Quel che resta. L'Italia dei paesi, tra abbandoni e ritorni. Donzelli Editore, Roma

Teti V (2018) Il sentimento dei luoghi, tra nostalgia e futuro, in De Rossi (a cura di) Riabitare l'Italia. In: Le aree interne tra abbandoni e riconquiste. Donzelli Editore, Roma

Tornatore JL (2011) L'inventaire comme deni de la recon- naissance. In Bortolotto, Chiara (sous la direction de), Le patrimoine culturel immatériel, enjeux d'une nouvelle catégorie. Editions MSH, Paris

Yourcenar M (1951) Mémories d'Hadrien suivi de Carnets de notes de Mémories d'Adrien. Librairie Plon, Paris

Zagato L (2014) Diversità culturale e protezione/salvaguardia del patrimonio culturale: dialogo (e contaminazione) tra strumenti giuridici. In Cataldi G, Grado V (a cura di), Diritto internazionale e pluralità delle culture. Editoriale scientifica, Napoli, pp 369–388

Zingari V (2015) Patrimoni vitali nel paesaggio. Note sull'immaterialità del patrimonio culturale alla luce delle Convenzioni internazionali. In Zagato L, Vecco M, a cura di. Citizens of Europe. Culture e diritti

Zoja L (2013) Utopie minimaliste: Un mondo più desiderabile anche senza eroi. Chiarelettere Editore, Milano

European Policies and Strategies Enhancing Projects of Slowness

Learning from Experience: A Set of European Policies

Federica Bianchi

Abstract Defining a dedicated national plan for soft mobility means creating a strategic tool for a country's growth. Indeed, investing in a large-scale network of cycle paths generates economic and social benefits, which also extend to other occupational fields that are not strictly related to cycling. They involve various actors and stakeholders, and offer an opportunity for cooperation between institutional bodies at various levels. An increasing number of European countries are defining a structured intervention strategy. Their experiences witness how this approach is proving to be essential to converge the many issues related to the project for a territorial infrastructure and the resulting effects. The purpose of analysing five case studies—The Netherlands, Germany, France, Denmark and Switzerland—is to reason on the many methods and tools available to promote cycling in an effective and coordinated manner, and to underscore certain factors, which are present in all experiences and are crucial for a successful strategy.

1 Why Choose a Set of Policies?

Many European countries have understood that investing in a cycle network has positive economic and social effects on the geographical areas involved. They have defined national strategies and tools required to establish goals and actions for the large-scale development of bicycle mobility, and for regional and municipal coordination. This is true for the urban cycling sector, and also for suburban and cycling tourism, which we shall particularly discuss in this paper.

Discussing cycling and pedestrian infrastructures on a large-scale means conceptualizing and intercepting also other occupational sectors, including planning for infrastructures, tourism, agricultural policies, sports and culture. Hence, the formulation of a national cycling strategy offers an opportunity for cooperation between the various bodies, which can be involved in such a large project.

F. Bianchi (✉)
Politecnico di Milano, Milan, Italy
e-mail: federica7.bianchi@mail.polimi.it

P. Pileri and R. Moscarelli (eds.), *Cycling & Walking for Regional Development*, Research for Development, https://doi.org/10.1007/978-3-030-44003-9_13

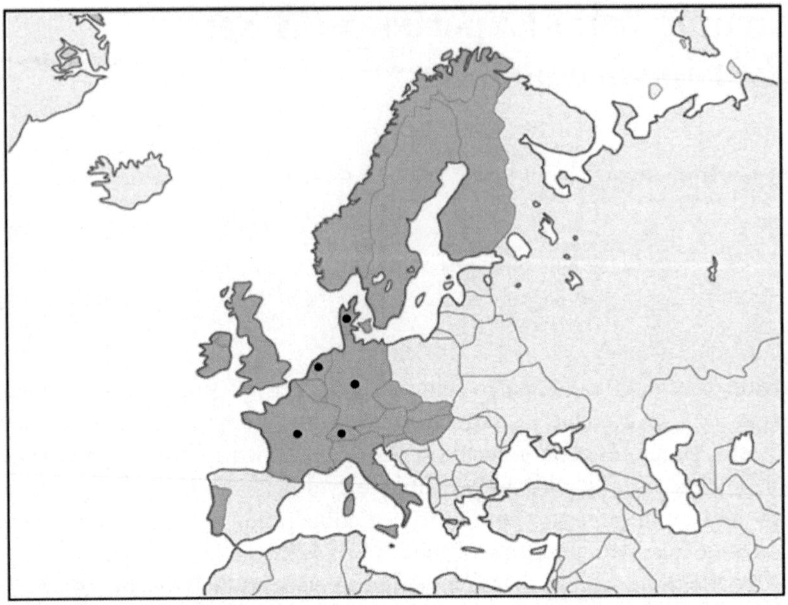

Fig. 1 Highlighted in dark grey, European countries with a national strategy for bicycle mobility. The dot marks the five cases discussed in the following chapters. *Source* the author

However, not all European countries have understood that these genuine slow tourism projects offer an opportunity for territorial regeneration. The need for a structured intervention has not always been understood (Fig. 1).

In this paper we shall describe the experiences of The Netherlands, Germany, France, Denmark and Switzerland. These five European countries have understood that having a national cycling strategy is essential to converge the many intrinsic aspects of a project, such as that of a territorial infrastructure. Based on this concept, they have defined action plans, established priorities for interventions and allocated funds for implementation.

The cases collected show us that defining national strategies and plans for cycle mobility is a crucial tool on which several long-term policies can be established in various sectors (e.g., tourism, economic development, landscape protection, etc.). The plans analysed concern various aspects of the extensive theme of cycle mobility. In each document, policies concerning suburban cycling tourism are included to a variable extent: they have been selected and analysed below. The comparison of these five different cases is intended to offer an idea to reason on the many approaches and tools that can be implemented to support territorial cycling in both an effective and coordinated way on multiple levels.

The same can be developed for walking routes and hiking trails. They too concern soft mobility and are potentially capable of generating miscellaneous beneficial effects in the territories crossed. However, the planning of walking routes is fragmented and presents several deficiencies in Europe today. The general delay in defining strategic visions, compared to what occurred for cycling, makes this form of mobility a sector that is still scarcely explored.

However, we report some important exceptions in the management of walking routes. Switzerland has a network of 7 hiking trails, with an overall length of more than 3,000 km, which cover the entire national surface area. They are managed by *SvizzeraMobile,*[1] which also handles cycling. The Netherlands present a network of 11,500 km of long-distance trekking routes (*LAWs—Lange-Afstand-Wandelpaden*[2]) featuring a large number of connections and dedicated road signs. Finally, France has a *Fédération Française de la Randonnée Pédestre,*[3] which manages the creation and signposting of 55,000 km of *Sentiers de grande randonnée.* None of the countries analysed has a strategic plan in place as yet.

2 Cycling in Europe: 5 Cases

The analysis of the five cases, which will be presented in chronological order from the first adopted strategy, focuses on reading the plans drawn up by the various institutional bodies to encourage bicycle mobility, particularly underscoring the theme of tourism. We shall note that some of them consider cycling as an extension of comprehensive politics about different sectors (e.g., Switzerland, *Sustainable development strategy 2016–2019*); in other cases, the main focus is on the bicycle, observed from various angles (e.g., Denmark, *Denmark—on your bike! The national bicycle strategy*). Some countries, like Germany and The Netherlands, have pioneered the integration of cycling into the mobility system, while others have adopted this systemic vision in more recent times (France and Switzerland).

For each tool, in the first place, we have related the antecedents that paved the way for the first strategic policy for cycle tourism, highlighting the many institutions that have promoted and developed it, and those involved in the subsequent phases. Then focus has shifted to the implementation modes of the strategy, the objectives, the actions envisaged and the funds allocated. Finally, the current status of plans has been compared against implementation of the network and future developments, if any.

2.1 *The Netherlands—Pioneers of Cycling*

The Netherlands is widely acknowledged as the country where cycling is most developed, both in Europe and in the world. Cities, such as Amsterdam, Eindhoven, Enschede and Utrecht, evoke scenarios related to a long-standing tradition in this

[1] www.schweizmobil.ch.

[2] www.wandelnet.nl/lange-afstand-wandelpaden.

[3] www.ffrandonnee.fr.

framework, but the great success of soft mobility cannot be simply archived as a 'cultural factor'. The fact that cycling here is so widely practised, becoming part of the country's cultural DNA, should lead us to muse over its origins, since cycling, which is widely acknowledged, is not the same as a few decades ago. In fact, until the mid-1970s the Government paid very little attention to soft mobility, merely regulating it, rather than incentivizing it. A unique circumstance, namely the combination of the energy crisis and the increasing mortality rates of pedestrians and of cyclists due to cars (with the subsequent creation of very active popular committees), caused the moderation of traffic and incentivization of alternative transport to become highly debated themes, which led to the first strategies to promote the use of the bicycle in order to reduce that of motorized vehicles.

The reference national strategy was formulated in 1990, the *Masterplan Fiets*, which regulated both urban and suburban cycling as an intrinsic part of a wider ranging transport policy designed to limit problems associated with mobility (e.g., low safety standards, consumption of fossil raw materials, traffic congestion, etc.).

The Masterplan was promoted by the Ministry of Transport, Public Works and Water Resources, and developed within the Transport Administration, also consulting the Dutch Union of Cyclists (ENFB, *Enige Echte Nederlandse Fietsersbond*[4]).

Several actors were involved during the strategy's effective period 1992–1998, including provincial administrations, municipalities and associations, thus acknowledging the multiscalar and multisectoral value of bicycle mobility, also with the intention of creating a broad base for policies and implementation of the activities envisaged.

The *Masterplan Fiets* focused on diminishing the use of the automobile in favour of the bicycle and public transport, reducing the use of non-renewable resources and soil consumption for new infrastructures. To obtain this, authorities admitted that cycling should be made as advantageous as possible, also making things more difficult for motorized mobility. In this process, communication and sensitization activities were deemed important for residents to understand and accept the argumentations of the strategy.

Five long-term objectives were defined, even reaching beyond the strategy's timeline to better optimize strengths in view of the future (Table 1).

Generally, the long-term objectives implied the implementation of services, of relationships and of organization. Single innovative projects developed in the short-term had the purpose of structuring policies at various administrative levels. As reported in the table, a tangible and verifiable numerical result was stated for each objective (except for safety) and it had to be achieved within an established timeframe.

The actions implemented concerned limiting access to cars in cities and in areas of natural interest, increasing the cost of motorized mobility, incentivizing alternative means of transport and cooperation between institutions, associations and residents. Particular attention was especially given to the role of research by assigning funds for the study and implementation of pilot projects, the formulation of long-term development models, education, communication and monitoring.

[4]www.fietsersbond.nl.

Table 1 Objectives established by the *Masterplan Fiets* in 1990. They were courageous goals that were ahead of their times, considering the period when they were formulated

Switching to the bicycle	An increase of 3.5 billion kilometres travelled by bicycle from 1986 to 2010[a], reducing the increase in the use of cars by ca. 9%
Switching to public transport + bicycle	Increase of 1.5 billion kilometres travelled by train from 1990 to 2010 (+15%) with an improvement in the intermodal system
Increased safety	15% reduction in fatal bicycle accidents from 1986 to 1995, and 50% within 2010. 10% reduction in accidents by 1995, and 40% by 2010
Preventing bicycle theft	Considerable reduction in thefts by the year 2000. A minimum result to be achieved is not specified for this objective
Developing communication	Integration of policies concerning cycling into the national, regional and local transport plan by 1995. Development of pilot models and projects

Source Directorate-General for Passenger Transport, 1999 «The Dutch Bicycle Master Plan. Description and evaluation in an historical context»
[a]Which every year correspond to ca. 10 km/resident travelled by bicycle

The first funds allocated during the period 1991–1995 amounted to 120 mln € (mean, 24 mln €/year)[5] for the creation of cycle paths, bridges, underpasses, parking lots for bicycles and other services for cycling.

The role of the State consisted in catalysing processes by focusing on an integrated approach, stimulating development, gathering information, supporting projects and establishing rules. Briefly, the State had to guarantee basic favourable conditions on which the Provincial Administrations and municipalities, as executors of the national strategy, could organize their work. Instead, regarding funds for research and pilot projects, the same decentralization process did not occur in order to avoid the risk of bad management and resource dissipation.[6]

The 1990 *Masterplan Fiets* is the most recent national strategy for cycling in The Netherlands. Since then increasing autonomy has been given to the 12 provinces, which are the country's second administrative order. The *Masterplan Fiets* remains a landmark, a clearly defined and practical document to establish provincial policies. The State maintains a facilitating role by supervising processes, coordinating regulatory aspects, and drawing the attention of other ministries concerned by the consequences of bicycle mobility developed at a territorial level.

The reference organization for the cycling network at a national level is *Landelijk Fietsplatform*: from 1987 it has collected and coordinated all institutions involved in cycling tourism. *Fietsplatform*, initially established as an independent entity, is now supported by its own members and by the 12 provinces to carry out two main duties:

[5]Directorate-General for Passenger Transport (1999) «The Dutch Bicycle Master Plan. Description and evaluation in an historical contex».

[6]Ibid.

(1) create a reference and support centre by publishing guidelines and information sheets for administrators; (2) coordinate the national cycling network by managing route development, maintenance and communication.

Fietsplatform's priority is to develop a network accompanied by appropriate road signs. The necessary criteria were defined before tracing the routes:

- link tourist centres, rural areas and cities;
- use existing infrastructures;
- cross pleasant and varied landscapes;
- establish connections between routes at various levels;
- connect points of interest;
- connect the main railway transport stations.

With the cooperation of local institutions and following several field surveys on bicycle, the *LF-network (Landelijke Fietsroutes network)* was established in 1990.

The network is made up of 26 *LF-routes,* routes entirely signposted for a total of more than 4,500 km to cover the entire country, also linking up to European cycle paths (Fig. 2). About 50 regional cycle paths, which form a capillary network of more than 30,000 km in the territory must be added to these. An operation is currently (2019) in progress to reduce the number of national cycle paths to 10 high-quality theme routes by 2021, privileging strategic segments and mutually combining several pathways.[7]

2.2 Germany—A Thick Cycling Network

Germany was one of the first European countries to develop a national strategy to support cycling. The process commenced already in the 1970s–1980s, initially with a pilot project dedicated to bike-friendly cities and later with a plan for the creation of cycle paths along the main federal roads. Particularly, the latter project was implemented by using part of the budget allocated to 'Renewal, restoration, improvement and construction of federal highways'. In 2002 it led to establish a separate budget only for cycle paths along highways.[8]

In 2000 the Federal Government, through the *Federal Ministry of Transport, Building and Housing*, took up the role of facilitator and coordinator in this sector, and presented the *National Cycling Plan 2002–2012 'Ride your bike! Measures to Promote Cycling in Germany*'. It is the first report for a national strategy in Germany, and was drawn up by a division that managed public transport and cycling at the Ministry. The strategy's established goal was to underscore the opportunities offered by developing cycling to promote an integrated transport policy, exploiting its potential in favour of more sustainable travel, over a period of 10 years. Precisely,

[7] https://www.holland-cycling.com/blog/274-lf-routes-to-be-overhauled.

[8] Federal Ministry of Transport, Building and Housing (2002) «National Cycling Plan 2002–2012. Ride your bike! Measures to Promote Cycling in Germany».

Fig. 2 Dutch national cycle network. Red, the 26 *LF-routes*. Today they have all been implemented (4,500 km/26 routes). *Source Stichting Landelijk Fietsplatform*

the table (Table 2) reports the seven themes defined by the strategy, each translated into objectives to be achieved.

To achieve the goals, the strategy defined a series of actions involving various actors and their respective competencies in multiple frameworks that could benefit by strengthening bicycle mobility. Hence, cycling was conceived as a system and was, therefore, planned as such. Broadly speaking, the actions envisaged planning and implementation of new infrastructures, connections, road signs and services; the construction of a new reference institutional framework; communication and education; research, monitoring and management.

Specifically considering cycle tourism, the objectives concerned achieving both environmental and economic effects. On the one hand it focused on reducing the

Table 2 Action plan for cycle tourism established by Germany in the *National Cycling Plan 2002–2012*

Daily use in the urban setting	Facilitate the use of bicycles by removing obstacles to their daily use
Cycle tourism	Incentivize the use of the bicycle even for occasional long-distance travel
Intermodality	Strengthen and implement connections between the various means of transport
Coordination, planning and promotion	Implement partnerships between entities and public authorities at various levels
Services supporting cycling	Improve the quality of infrastructures and draw up an interconnection plan to link up the national and regional network
Legislative framework	Review safety standards and attractive features of cycling infrastructures, and review freedom of action at a local level
Safety	Increase the safety of all users, especially those most at risk (children and the elderly)

Source the author, starting from Federal Ministry of Transport, Building and Housing, 2002 «National Cycling Plan 2002-2012. Ride your bike! Measures to Promote Cycling in Germany»

production of pollutants and the use of non-renewable energies, even in the tourist sector, by enhancing the environmental awareness of both German and European residents. On the other hand it mentioned all sectors that would benefit by the growth of cycle tourism, such as, for instance, the bicycle industry, services and hospitality sectors: the estimated turnover was 5 billion €/year.

To achieve these goals, it was deemed necessary to plan cycling as a convenient and alluring alternative, which would attract the attention of a broad user basin, not only of sportsmen and enthusiasts. Such targeted action was proposed by improving the overall tourist offer. Each action was assigned to competent actors with precise indications concerning the measures to be adopted and the extent of the interventions to be implemented. As reported in the table (Table 3), the German approach basically considered cooperation, at various levels, between Federal States, local administrators, tourist entities, companies and associations. Particularly, coordinating actions were carried out nationally, while the more specific and operational ones were the responsibility of local entities.

This plan of policies and actions presented the *D-network*, the national cycle network, whose 12 paths cover the entire surface of the country, attempting to harness part of the existing regional routes that meet safety, beauty and functional requirements. In the framework of the 2002 plan, a 420 km pilot project was developed on the *D-Route 12 Oder-Neisse-Radweg*, with economic support from the *Federal Ministry of the Economy and Technology* (Fig. 3). The project consisted of implementing this significant segment as an opportunity to propose management and development models, which could later be transferred to other *D-Routes*. The choice of this segment along the Polish border, crossing marginalized areas of Eastern Europe, was intended

Table 3 Action plan for cycling tourism established by Germany in the *National Cycling Plan 2002–2012*

Theme areas	Actions	Actors	Interventions
Quality of the infrastructure	Create convenient, safe and attractive cycle paths; resting areas with seats and shelter	Local authorities, construction companies, tourist entities	12 national cycling routes (10,200 km) and regional routes (30,000 km); 4,000 picnic areas and 2,000 resting places with shelter
Road signs	Create destination signs in compliance with the guidelines of the German road code (FGSV); create a common national identification logo	Federal States, construction companies, local authorities, *German National Tourist Office*, *German National Tourist Board*, *German Cyclists' Federation*	National system of cycle routes for tourists (ca. 10,200 km); pilot project of the Oder-Neisse-Radweg cycle route
Management and maintenance	Regularly check the infrastructure and road signs	Tourist entities, construction companies, local authorities	Monitor the entire network of more than 40,000 km
Hospitality	Open new hospitality facilities near the system	German Cyclists' Federation, tourist entities, companies operating in the hospitality sector	2,900 new bed&bike inns

Source Federal Ministry of Transport, Building and Housing, 2002 «National Cycling Plan 2002–2012. Ride your bike! Measures to Promote Cycling in Germany»

to achieve economic effects resulting from the presence of such an infrastructure by incentivizing the development of local economies.

When the Plan was launched in 2002, for the first time the Federal Government decided to allocate dedicated funds, amounting to 100 mln €, for the creation and maintenance of cycle routes along roads within federal competence. It was an initial investment, which referred only to the year in progress. However, it was meant to give a strong example to other entities related to bicycle mobility, encouraging them to make a commitment within their fields of competence.[9]

An additional form of funding was provided by the *GVFG—Local Authority Transport Infrastructure Financing Act*, which provided the Federal States with funds from the mineral oil tax. These funds, which amounted to a total of 1.7 billion €/year, could be used to implement the general infrastructural system of municipalities (roads, public transport lines, decommissioned railway segments, etc.), also including cycle paths, road signs and parking lots for bicycles near public transport

[9]Ibid.

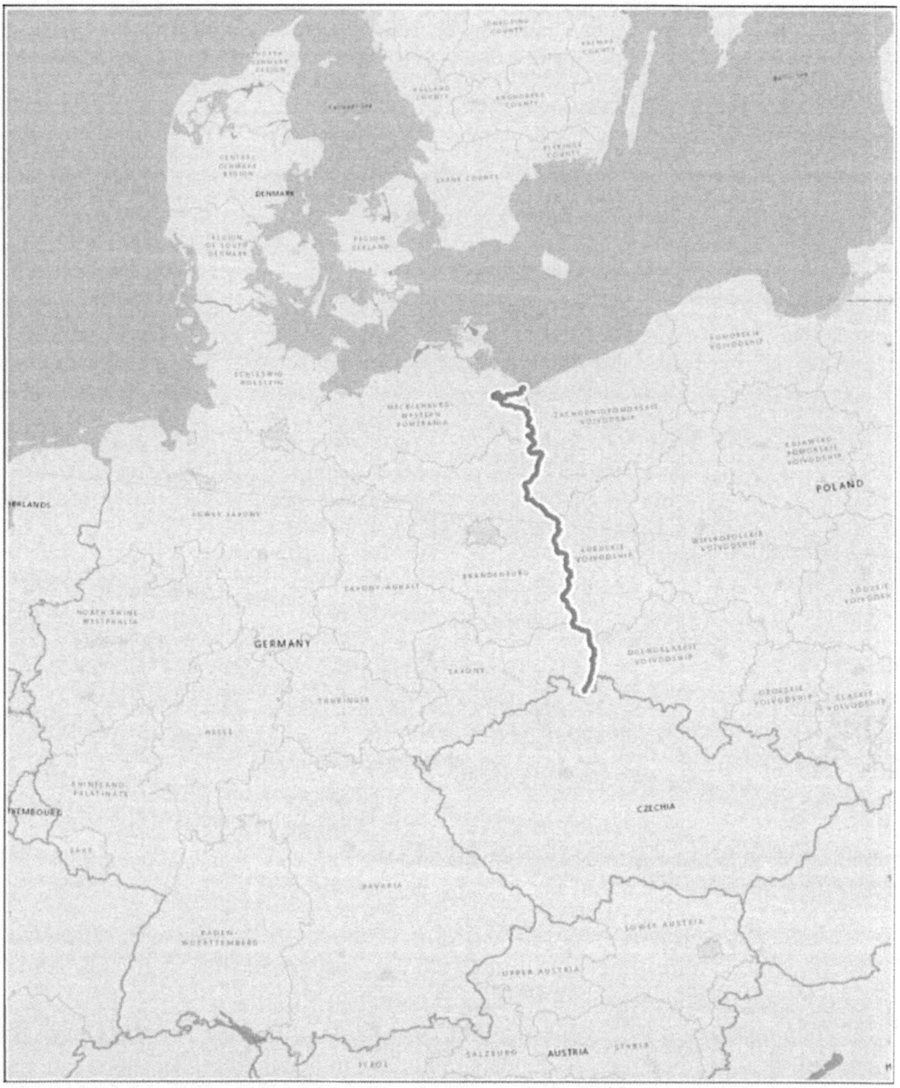

Fig. 3 Map of the pilot project *D-Route 12 Oder-Neisse-Radweg*, which runs along the national Polish border. *Source* the author

hubs. Though the funds were not exclusively dedicated to bicycle mobility, it must be said that the Federal Government recommended the States to prioritize interventions supporting cycling.

The plan currently in force—*2020 National Cycling Plan—Joining forces to evolve cycling* was launched at the end of the planning period 2002–2012. The Plan is in line with the previous edition. It is not considered an update but rather an evolution with a strategy focused on a participatory process involving a wider range of stakeholders from the Federal States, the local authorities and the academic world.

Table 4 Actions concerning cycle tourism established by the *2020 National Cycling Plan,* based on the various responsibility levels

Federal Government	Support implementation and expansion of the *German Cycling Network*; standardize and share databases with planners; supervise and evaluate; provide consulting services for funding opportunities
Federal States	Coordinate the development of the cycle tourism offer; provide consulting services for funding opportunities; coordinate integrated planning of bicycle mobility and public transport
Regional Administrations and local authorities	Deliver funds for maintenance, quality improvements and services for cycle tourism
Stakeholders	Evaluate and monitor quality standards

Source the author, starting from Federal Ministry of Transport, Building and Urban Development, 2012 «National Cycling Plan 2020. Joining forces to evolve cycling»

The strategy was promoted by the *Federal Ministry of Transport, Building and Urban Development*, defined and coordinated by the *Passenger Transport, Public Transport System, Cycling* division. The reference period has diminished to 8 years (2012–2020), compared to the previous edition. The goal is to reinforce the vision of bicycle mobility as a system made up of three equally important elements to be developed in a balanced manner: (1) infrastructure, (2) promotion and communication, (3) supporting services.

The plan defines 9 theme areas, and the main novelties, compared to the 2002 plan, concern electric pedal-assisted cycles and an improvement in infrastructural quality. Even the section on cycle tourism follows the previous plan, with special focus on the aspects of coordination and dialogue between the Federal Government and local authorities. It is also highlighted that, to achieve the goals, it does not suffice to have an increasing larger network; instead, it is necessary to focus on improving the quality of existing infrastructures and services.

The actions relative to cycle tourism concern various levels of administrative responsibility: Federal Government, Federal States, Regional Administrations/local authorities, stakeholders (Table 4). As with the previous version of the National Plan, there is once again a synergy between actors and their competencies, with duties concerning national coordination and local operational management.

Today (2019) the 12 *D-Routes*[10] cover a length of 11,700 km; the first 6 routes are placed in a West to East direction, while the remaining travel from North to South, linking all the German regions[11] (Fig. 4). As mentioned above, considerable

[10]www.radnetz-deutschland.de.

[11]This paper only refers to 12 routes that make up the national network. There are also other main cycle tourism routes, amounting to a total of 50,000 km, that do not belong to the D-Routes (source: BMWi, 2009, «Grundlagenuntersuchung Fahrradtourismus in Deutschland», Forschungsbericht no. 583).

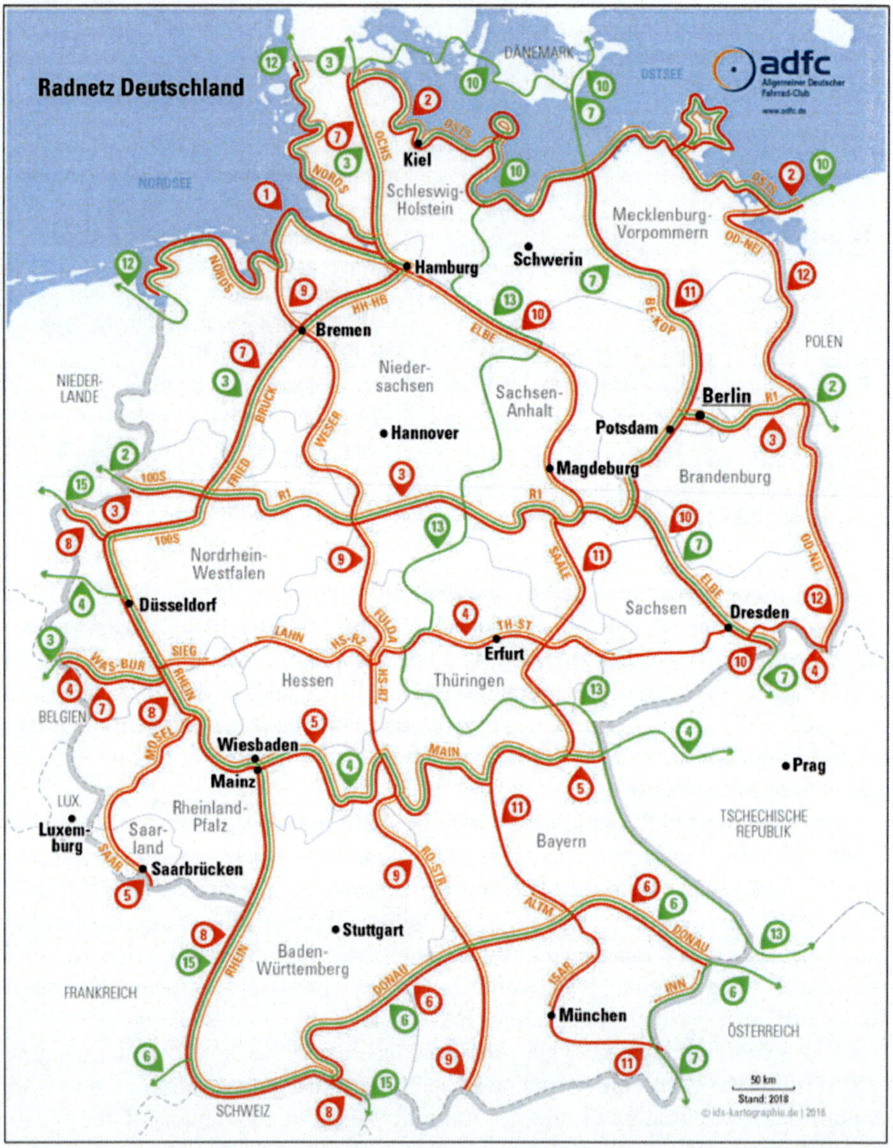

Fig. 4 The German national cycle network. Red, the 12 *D-Routes* (11,700 km). *Source Allgemeiner Deutscher Fahrrad-Club*

importance is given to the quality of the paths in order to establish a basic standard and to improve it in the course of time. These paths mostly run beside waterways and decommissioned railway tracks, or they are theme-based, organized according to the territories they cross. Made up of existing cycle paths, they have been chosen to connect starting points and strategic destinations in a safe and comfortable manner (Table 5).

Table 5 Overview of the 12 German *D-Routes* and of their respective essential features

D-Route	Length[a] (km)	Description
D-Route 1 *Nordseeküstenroute* (North Sea route)	907	Runs along the coast from The Netherlands to Denmark, through the estuaries of the rivers Ems, Weser and Elbe, exploring mercantile shipping
D-Route 2 *Ostseeküstenroute* (Baltic Sea route)	1055	Along the Baltic coast, it touches many seaside resorts, coastal landscapes and historical cities, crossing certain hillside landscapes
D-Route 3 *Vreden-Zwillbrock/Küstrin-Kietz*	960	Crosses the country from Vreden (at the Netherlands border) to Küstrin-Kietz (Polish border), traversing a rich variety of continuously changing landscapes
D-Route 4 *Mittelland-Route* (Internal route)	1045	Follows the woodland strip, which crosses Germany from Aachen (Belgium border) to Zittau (Polish border), making its way along low mountain ranges
D-Route 5 *Saar-Mosel-Main*	1021	Crosses various vineyards from West to East, exploring local wines from Saarbrücken (French border) up to the district of Upper Franconia (Czech Republic border)
D-Route 6 *Donauroute* (Danube route)	733	From the Swiss border near Basel, it crosses Swabia and Bavaria along the Danube, up to Passau (Austrian border)
D-Route 7 *Pilgerroute* (Pilgrim's route)	1189	Follows the trail to Santiago de Compostela from Flensburg (Danish border) to Aachen (Belgium border)
D-Route 8 *Rhein-Route* (Rhine route)	1019	Follows the German sector of the Rhine from the Netherlands border to Switzerland, crossing the historical industrial cities that developed as a result of the presence of this important shipping waterway
D-Route 9 *Weser—Romantische Straße* (Romantic route)	1288	The most direct North–South line, it crosses the country from Bremerhaven on the North Sea to Füssen (Austria border), featuring a variety of historical and cultural monuments
D-Route 10 *Elberadweg* (Elba route)	1328	Follows the course of the Elbe, from the estuary in Hamburg, on the North Sea, to the Czech Republic border, crossing large areas of alluvial plain

(continued)

Table 5 (continued)

D-Route	Length[a] (km)	Description
D-Route 11 *Ostsee—Oberbayern* (Baltic Sea—Upper Bavaria)	1697	Germany's second largest North–South crossing (besides D-Route 9), it starts from Rostock, on the Baltic Sea, and reaches the Austrian border, near the city of Salzburg. It crosses plains and low mountain ranges covered with woodland
D-Route 12 *Oder-Neiße-Radweg*	630	Runs along the eastern border of the country, crossing marginalized regions near Poland and intercepting the Sorb settlement of Slavic Sorbian-speaking people, the only ethnic minority in Germany

Source the author

[a]Data, obtained from de.wikipedia.org/wiki/D-Route, contain some inaccuracies, probably due to overlapping segments and to different calculation methods adopted by the sources. The total length of the network is 11,700 km (www.radnetz-deutschland.de/radnetz-deutschland), while the sum of the individual D-Routes is 12,800 km

The plan also regulates the allocation of funds for cycling, particularly for the creation of new infrastructures, for maintenance or to renovate the existing ones, besides service implementation. As occurred with the 2002 plan, the Federal Government provided the necessary investments for the cycle paths along federal roads, both new and existing ones. To quantify allocations, the Federal Government, along with the Federal States and local authorities, will conduct studies to determine the need for funds at various levels for the plan's entire effective period. However, an initial estimate is submitted, including creation and maintenance of cycle paths, parking lots for bicycles, communication, sharing stations and other similar services. The average need for funds calculated is 13 €/resident per year, considering that the value can vary depending on geographical area, initial situation and future prospects.[12]

The plan also introduces a federal financial assistance plan of 3.2 mln €/year to support projects that do not envisage investments, and which are highly innovative and replicable in different settings. More than 180 projects have been funded at present (2019) to improve sustainable transport.[13]

The prospects of the *National Cycling Plan* are not limited to the close of 2020 but are, instead, conceived for a longer period of time. Indeed, the strategy is studied to make the most of potential bicycle mobility by looking to 2050, when the bicycle will be even more extensively used. It is necessary to already start implementing the system and integrating cycling into the scene of mobility.

2.3 France—A Recent History

France has a more recent story of cycle mobility. A national network was first discussed in 2001, defining some basic guidelines in the *Cahier des charges – Schéma National des Véloroutes et Voies Vertes.*[14] The document was drawn up by the French Tourism Engineering Agency and sponsored by the Ministry for Territorial and Environmental Management, the Ministry of Transport, the Ministry of Youth and Sports and the State Secretariat for Tourism. It is clearly a technical document that is both concise and complete.

It proposed creating a network of long-distance cycling routes to meet the expectations of both the French and the European population. Identified as a tool for territorial planning, which would also enable the development of a long-lasting and sustainable tourist offer, the network was constructed following 5 guiding principles:

- create a 7,000–9,000 km cycle route system;
- use existing and easily accessible infrastructures;
- link up the leading cities;
- ensure the presence of at least one route per region;

[12]Federal Ministry of Transport, Building and Urban Development (2012), «National Cycling Plan 2020. Joining forces to evolve cycling».

[13]www.nationaler-radverkehrsplan.de/de/bund/foerderprogramm.

[14]www.af3v.org/-Cahier-des-charges-national.

- ensure continuity with the networks of bordering countries.

Moreover, each route in the network had to possess certain characteristics specified in the document. Particularly, each route had to be:

- Linear: allow to reach a place with the most direct route;
- Continuous: without interruptions, even when crossing town centres;
- Safe: not generate dangerous situations, for instance, with motorized traffic, near river banks or bridges;
- Signposted: have uniform, clear and exhaustive indications;
- Be maintained: always be accessible for travel in every segment;
- Suitable for all: be usable even by inexpert cyclists (e.g., the roadbed must not be bumpy, minor inclinations within a 3% threshold, etc.);
- Supported by other services: e.g., hospitality, transports, promotional initiatives.

In 2011 the former Minister of Transport Thierry Mariani established the Work Group for the development of bicycle mobility (*Groupe de travail pour le développement de l'usage du vélo*), made up of representatives of associations and constructors related to the bicycle scene, and of representatives of other ministries. In 2012 the Work Group presented the *Plan National Vélo* on behalf of the Ministry of Transport. This document studied measures to incentivize the development of both urban and suburban cycling, based on the experience of other countries (including The Netherlands, Germany and Denmark) to encourage practical proposals and projects.

The strategy's first objective was to achieve a mean increase of +1% a year in bicycle use, reaching a 10% modal share of the general volume of traffic by 2020. This objective is associated with other purposes:

- take action against climate change;
- develop policies favouring health and to control public health-related expenditure;
- increase the autonomous movement of people.

The report proposes a series of actions to achieve these objectives. They can be grouped into 6 themes:

- infrastructure;
- urban cycling;
- tourism and sports;
- value enhancement of economic, social and environmental aspects;
- bicycle;
- role of the State.

As regards tourism, the *Plan Vélo*'s specific goals were conceived to make France the first destination in Europe for cycle tourism, achieving economic benefits (especially locally) from the creation of paths and accessory cycle route services.

The table (Table 6) reports the actions defined to achieve these goals, which concern the creation of a national cycle route network, dissemination of good practices, research, management and monitoring. The plan neither establishes the

Table 6 Action plan for cycle tourism defined by the *Plan National Vélo*

Create the *SN3V*, *Schéma National des Véloroutes et Voies Vertes*	Integration with the *Schéma National des Infrastructures de Transport*; facilitation of the possibility of obtaining financial contributions from the European Union; single management of road signs as a trace for the future; extension and facilitation of the contribution of the State's public institutions; development of partnerships with *France Vélo Tourisme*; adoption and update of *Schémas Régionaux de Véloroutes et Voies Vertes*, with regular regional committee meetings
Enhance the value of sports	Enhance the value of iconic manifestations; open cycle training schools; create cycle paths for sports purposes
Favour the State's role in the creation of reports	Provide support for environmental and landscape themes, risk management and environmental restoration
Update knowledge of the cycle tourism market	Study the foreign demand and offer of competing countries
Support *the ON3V*, *Observatoire National des Véloroutes et Voies Vertes*	Create an observatory with the Association of Departments and of Cycling Regions; make a GIS database on cycling routes and greenways available
Consolidate the *MN3V*, *Mission Nationale des Véloroutes et Voies Vertes*	Coordinate the national development policy for the *SN3V*; facilitate creation of the system

Source the author, starting from Groupe de travail pour le développement de l'usage du vélo, 2012 «Plan National Vélo»

methods nor the entity of financial support required for the implementation of these actions.

The Ministry of Transport, Sea and Fisheries and the Ministry of Ecology, Development and Energy published the *PAMA—Plan d'action mobilités actives* in 2014. The document, drawn up by a division of the Ministry of Transport, presented the optimal conditions to implement an ambitious, wide-ranging vision concerning active mobility. As a follow-up to the 2012 plan, the topics are discussed with 6 themes:

- intermodality of public transport and soft mobility;
- protection of public space and safety for soft mobility;
- value enhancement of the bicycle's economic benefits;
- integration of soft mobility policies into urban planning;
- development of leisure and cycle tourism routes;
- communication and dissemination of the bicycle market's beneficial effects.

It has been observed that the most interesting cycle tourism axes were axis 1 and axis 5.

Axis 1 focuses on the development of intermodality as the integration of long-distance travel, ensuring continuity and complementarity between various types of mobility. Intermodality is conceived both in terms of daily travel and for tourism.

Axis 5 is dedicated to the development of cycle tourism, seeking to meet the so-called 'concrete social demand': cycle tourism embodies the values with which people increasingly identify themselves, namely authenticity, freedom, well-being and respect for nature. Hence, long-distance cycle routes are considered tools to revive rural areas and the bearers of economic activities. The objective to be achieved is to increase the percentage of cycle tourism holidays by 3% compared to traditional tourism, reaching 6% by 2020. This increase in cycle tourism is estimated to generate 12,000 new jobs and profits amounting to 2 billion €.

Finally, the new *Plan Vélo* was introduced in 2018. It featured no innovations on the theme of cycle tourism, compared to the previous edition; therefore, the 2012 edition and technical documents relative to 2014 are still referred to today.

The national cycle network is made up of 47 routes with a total of 20,000 km. These include 10 main ones with a total length of 5,116 km (Fig. 5). The 10 main routes were chosen based on (1) length longer than 300 km, (2) existing for at least 50% and (3) in their own site for up to at least 20%.

2.4 Denmark—Strengthen to Improve

Denmark is one of the most developed countries in Europe for bicycle mobility. The bicycle is now an intrinsic part of its overall transport system. However, a strong tradition does not suffice to maintain high standards on the topic of cycling. This is evidenced by fact that the use of the bicycle is declining even in an avant-garde country, such as Denmark: 2013 recorded a more than 10% drop in the number of cyclists in the country, compared to 1990.[15]

In 2014 the Ministry of Transport, in partnership with the various organizations and municipalities, drew up and published *Denmark – on your bike! The national bicycle strategy*, a national strategy studied to make bicycle mobility more convenient and accessible by promoting its daily use and also cycling for leisure and tourism.

As reported in the table (Table 7), the 'three pillars' that have defined the strategy are linked to an equal number of specific actions supporting the increased use of the bicycle.

Particularly, concerning cycle tourism, the need for a strong infrastructural investment is stated in order to have steadily improved and well maintained routes, which can be easily travelled by everybody. Moreover, each route must be carefully located in a strategic and accessible position. Specific actions conducted by the Ministry of Transport include improving the national signposting system, establishing a study group for the development of cycle tourism, planning new cycling tourism routes and creating an interconnected and accessible network.

[15]Transportministeriet (2014), «Denmark—on your bike! The national bicycle strategy».

Fig. 5 French national cycle network. The segments created today account for almost 50% of the total extension (2,300 km out of 5,116 km). *Source Vélo and Territoires*

However, cycling tourism does not only concern the transports sector but concomitantly involves various institutions and organizations: a broad partnership base ensures the development of a uniform, high-quality network. Hence, the infrastructural project must be implemented in line with innovations in the accessory services system, constantly monitoring progress, with a communication project and with research.

Currently (2019) there are more than 12,000 km of signposted routes in Denmark, 4,325 km of which make up the national network's 11 routes (Fig. 6). The municipalities are responsible for planning the paths, which must be approved by the *Danish Road Directorate*, the agency responsible for state roads in Denmark, to obtain a consistent and coordinated network also based on regional and national development. Moreover, the network must be created by taking into account several established criteria (Table 8).

Table 7 The three themes, and the relative actions, on which the strategy is based

Daily cycling	Combine bicycle use and public transport; provide high-level parking lots for bicycles; provide solutions for companies to incentivize cycling; create continuous cycling routes; support research for innovative solutions
Cycle tourism and entertainment	Provide high-quality infrastructures for cycle tourism; plan a consistent signposting system; experiment with promotional initiatives to involve people; define an adequate hospitality offer
Safety and beginner cyclists	Create suitable routes for everybody with high safety standards; develop programmes in schools; promote campaigns addressing an extensive public of inexpert cyclists; create projects for accident prevention

Source the author, starting from Transportministeriet, 2014 «Denmark—on your bike! The national bicycle strategy»

For 2014 the Government allocated funds amounting to 57 mln € (426 mln DKK), including 24 mln € for the creation of cycling highways and parking lots for bicycles, 23 mln € for new national and local works, 6.6 mln € for research and implementation of new project solutions, and 2.8 mln € for accident prevention.

Future network expansions must be considered from time to time to ensure strategic planning that is consistent with the existing status.

2.5 Switzerland—Cycling for Sustainability

In the European scene, Switzerland is a particular case for national planning of the cycle network, which is managed by a subject outside the Government, with the cooperation of the Swiss Cantons and of the Swiss Tourism Federation.

The reference strategy for bicycle mobility is, instead, the *Sustainable Development Strategy 2016–2019* defined by the *Federal Office for Spatial Development—Sustainable Development Section* on behalf of the *Swiss Federal Council*. It is a wide-ranging document, which establishes priorities for the country's medium- and long-term sustainable development, covering several themes involved in achieving the objectives of the United Nation's 2030 Agenda. The strategy is based on the need for dialogue and cooperation, both between various administrative levels, and between associations, companies, organizations and research institutions, to produce a common action plan and a framework for the formulation of actions implementing this strategy.

The theme of bicycle mobility has been added to the strategy in the action area called *Urban development, mobility and infrastructure*. A general vision is first provided by presenting the objectives to be achieved. The long-term goals are (1) sustainable development of cities and infrastructures, preserving farmland and nature

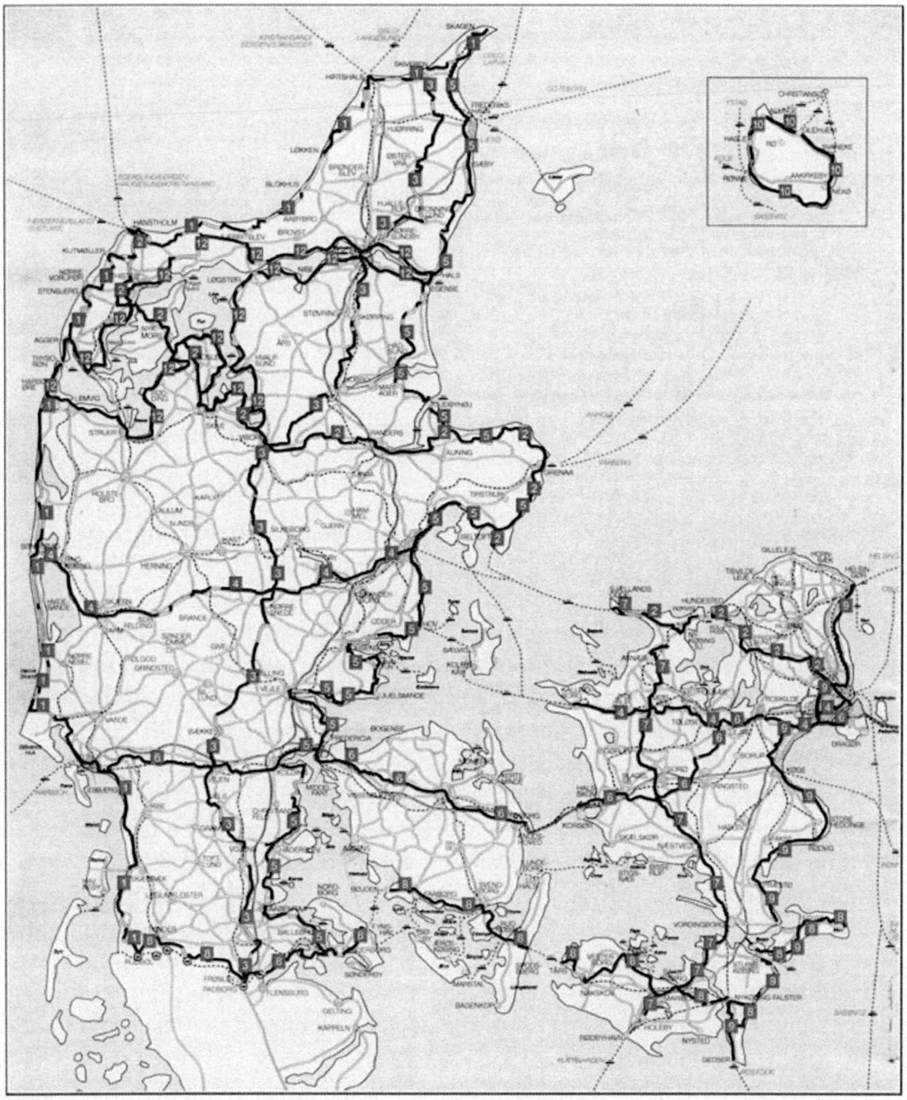

Fig. 6 Danish national cycle network. It has now been completed (4,325 km/11 routes). *Source Cycling Embassy of Denmark*

areas, and (2) a coordinated transport system, which includes various means of transport with balanced and synergistic development of a network and of multifunctional nodes. The medium-term goal, scheduled for 2030, is to reduce motorized traffic to levels that guarantee the population's safety and health, and the use of the bicycle on the part of all user categories throughout the nation.

Strategic actions on the theme of mobility are aimed at constructing an efficient, multisectoral system that is sustainable in economic and environmental terms.

Table 8 Criteria for the creation of the Danish national cycle network

Safety	Avoid cycle paths in heavy traffic roads and avoid them crossing an excessive number of junctions
Attractive features	A cycle path must be an added value, allowing to cross places with high cultural and natural value
Services	Consider the possibility of implementing the offer of tourist hospitality (accommodation, catering, trade, mechanical assistance, etc.)
Convenience	Avoid high inclinations, bumpy roadbeds, hairpin bends
Road signs	Establish a signposting system that is consistent with guidelines defined by the *Danish Road Directorate* and with regional and local routes
Meaning	A national route is a main road and must have a meaning (e.g., theme route, link to European routes, etc.)

Source the author, starting from Transportministeriet, 2014 «Denmark—on your bike! The national bicycle strategy»

The *Action plan for non-motorized transport* is currently (2019) being developed to improve the conditions of independent and multimodal bicycle mobility, thus making it part of a safe, accessible and attractive system.

Hence, the *Sustainable Development Strategy 2016-2019* only provides an overall vision and a series of objectives to be achieved, assigning the definition of more specific and targeted strategies to the lower levels.

The *Federal decree on cycle paths, footpaths and hiking trails* (13/03/2018), whose application is currently being processed, was approved in 2018. The decree establishes that the Confederation shall define principles concerning the cycle network, and may support and coordinate provisions laid down by the Cantons and by third parties in this regard. The Cantons, along with the Municipalities, will have competence on the subject in relation to the planning, implementation and maintenance of cycle paths. The Confederation's role will be limited to supporting provisions adopted by Cantons and Municipalities, to defining framework principles and conditions, to coordination and communication. This already occurs for footpaths and hiking trails, and the adoption of the decree will ensure that the same juridical rules apply to cycle routes.

Legally acknowledging the value of cycle routes is a strong sign for Cantons, Municipalities and the Confederation. It paves the way to assign the right importance to cycling and the related infrastructures, and to increase efforts made to support this form of transport. Practical tools will also be arranged for the management of infrastructures during every phase (e.g., manuals, practical guidelines, databases, etc.).

Finally, it must be said that adopting the decree does not imply any financial commitment on the part of the Confederation, neither does it interfere with Cantonal and Municipal sovereignty.

Today, planning cycle tourism networks is the competence of *SvizzeraMobile*, a platform promoted by *Fondazione SvizzeraMobile.* It was created in 2008 for the development of slow tourism. The foundation is supported by Federal Offices,

Cantons and the Principality of Liechtenstein. *SvizzeraMobile* was initially financed mostly by public funds, while today 50% of support is provided by federal/cantonal contributions, and 50% is given by private contributions (e.g., partners, hospitality activities, tourism boards, etc.). Besides planning, *SvizzeraMobile* manages the signposting system, coordinates the hospitality offer, ensures integration with other means of transport and communication, while also cooperating with Cantonal Offices for infrastructures, which are appointed to perform construction works and maintenance.

The national cycle network is made up of 9 routes with an overall length of 3,300 km (Fig. 7), integrated by 54 cantonal routes and 49 local routes.

Fig. 7 Swiss national cycle network. It has now been completed (3,300 km/9 routes). *Source SvizzeraMobile*

3 What We Have Learnt

We have considered countries that are very different in terms of surface area, number of residents, administration and tradition, with a complex picture of actors involved and policies implemented to promote cycling (Table 9).

Despite the many differences, the interpretation of the five case studies reveals certain recurrent elements, which run as a common thread through national strategies. All cases present a national control room, followed by regional and municipal ones. It can be said that, in most cases, the strategies implemented establish precise actions to develop cycling in the entire nation, after first defining specific goals and a general inspiring vision. The level of detail in the study of the various layers refers to the coordination of policies, to the exchange of good practices, to investments in infrastructures, to funds for pilot projects, to support for innovation, to research and communication, assigning the most detailed aspects relative to the implementation and maintenance of infrastructures to regional and local levels.

Each case has implemented strategies entailing different resources and timelines. The table (Table 9) shows the length of the national paths in relation to the surface area of the relevant country: The Netherlands and Denmark have a thicker network, with 0.1 km of national cycle routes for every km^2 of surface area, while the country with the scarcest network is France (0.007 km/km^2). This aspect is also influenced by the territorial morphology of each nation, for instance, by the presence of mountain ranges or of extensive flatland. Another factor compared is the mean length of cycle paths per resident, which reveals that the part of the national routes available to every resident in Denmark (0.8 m/resident) is far higher than in other countries (Switzerland: 0.4 m/resident; The Netherlands: 0.2 m/resident; Germany: 0.1 m/resident; France: 0.07 m/resident).

Instead, we were unable to compare data relative to funds provided for the implementation of the various plans. The Netherlands allocated funds to create infrastructures and relative services within a specific timeframe. Denmark only made the investment for the year in which the strategy was launched, without presenting future prospects. The German plan states certain possible financing methods for cycling. We have no data for France and Switzerland.

Two main models have been traced to coordinate the development of bicycle mobility at a national level: on the one hand the presence of a separate body (e.g., *SvizzeraMobile*), and on the other hand the establishment of a dedicated office in an existing governmental department (e.g., *Groupe de travail pour le développement de l'usage du vélo*). Both models have positive and negative aspects. A nationally funded governing body entirely dedicated to cycling with a specific mandate could be more autonomous and effective but, at the same time, cycling might go on being perceived as something that stands apart from the rest of the transport system, and thus be considered secondary. Instead, a dedicated office within an existing department can be established more rapidly, but it might also risk feeling the repercussions of changes in the government more intensely.

Table 9 Summary and comparison of the illustrated case studies, with particular focus on the national cycle network based on population and surface area of each country

	The Netherlands	Germany	France	Denmark	Switzerland
Reference strategy	Masterplan Fiets	2020 National Cycling Plan—Joining forces to evolve cycling	Plan National Vélo	Denmark—on your bike! The national bicycle strategy	Sustainable Development Strategy 2016-2019
Activating strategy	Masterplan Fiets	2002-2012 National Cycling Plan 'Ride your bike! Measures to Promote Cycling in Germany'	Cahier des charges—Schéma National des Véloroutes et Voies Vertes	Denmark—on your bike! The national bicycle strategy	Sustainable Development Strategy 2016–2019
Planning period	1990–1997	2013–2020	2018–2025	2014–N.A.	2016–2019
Sponsor	Ministry of Transport, Public Works and Water Resources	Federal Ministry of Transport, Building and Urban Development	Ministry of Transport	Ministry of Transport	Swiss Federal Council
Actors involved	Administration for Transport, Ministry of Agriculture, Provincial Administrations, Cyclists' Union	Federal Government, Federal States, Regional Administrations and local authorities, stakeholders	French Agency for Tourism and Environmental Engineering, Ministry of Youth and Sports, State Secretariat of Tourism	Central Government, Regional Administrations, local authorities	Federal Office for Spatial Development, SvizzeraMobile
Surface area	41,543 km^2	357,578 km^2	675,417 km^2	43,094 km^2	41,285 km^2
No. of residents	17.2 mln	82.3 mln	68.3 mln	5.7 mln	8.5 mln
No. of national paths	26	12	10	11	9
Length of national paths	4,500 km	11,700 km	5,116 km	4,325 km	3,300 km

(continued)

Table 9 (continued)

	The Netherlands	Germany	France	Denmark	Switzerland
Mean length of national paths	173 km	975 km	511 km	393 km	367 km
Mean length of paths per resident	0.2 m/resident	0.1 m/resident	0.07 m/resident	0.8 m/resident	0.4 m/resident
Density of paths	0.1 km/km^2	0.03 km/km^2	0.007 km/km^2	0.1 km/km^2	0.08 km/km^2

Source the author

Whatever the model, we observed that the following factors are crucial for the strategy's success:

- the presence of various administrative levels;
- the presence of various stakeholders, even in settings not closely related to bicycle mobility (administrations, associations, research institutions, commercial activities, etc.);
- the presence of scheduled funds (or funds unrelated to extraordinary events);
- research activities;
- definition (and application) of a dedicated regulatory framework.

All the examples illustrated present a national intention that, despite being developed in different ways, focuses, in any case, on developing bicycle mobility in a coordinated manner at various administrative levels. This is not happening as yet in Italy. Italy has neither a dedicated ministerial department nor an agency. However, the Ministry of Infrastructures and Transport, and the Ministry of Cultural Heritage and Activities and Tourism have recently supported national policies for soft mobility, responding to a collective request made by several initiatives launched by various subjects over the years. These policies also include the National Cycling Tourism System, a cycle network of ca. 20,000 km. Ten priority cycle routes have been identified (with an overall length of 5,700 km) that received dedicated funds, pursuant to the 2016 and 2017 Budget Law (see Fig. 3.1, Sect. 11).

A common vision is required, defined by exploiting a wide range of strategies, both 'hard', which concern the infrastructure, and 'soft', in the educational and communications sector. The presence of a plan and of a precise structural organization of the various actors' functions allow to converge the many different elements involved in planning and disseminating suburban cycling nationally. Indeed, a single common strategy would ensure coordination, which means maintaining an overall vision, implementing additional wide-ranging policies, making the most of efforts and investments, formulating consistent objectives and speeding up the processes to achieve them. For the theme of suburban bicycle mobility to gain central importance in the scene of national policies, these cases indicate the need to address all these aspects in a synergistic and integrated manner. Single isolated initiatives seem scarcely effective, compared to these organizational structures, which are more reliable, guarantee the outcome and trigger all the benefits a line is potentially capable of generating.

References

AA. VV (2014) «Plan d'actions pour les mobilités actives. La marche et le vélo»
AA. VV (2014) «Plan d'actions pour les mobilités actives. Réunion du comité de pilotage»
AA. VV (2001) «Schema National des Veloroutes et Voies Vertes. Cahier des charges»
Allgemeiner Deutscher Fahrrad-Club—adfc.de [Oct 2019]
Bundesamt für Strassen—astra.admin.ch [Sep 2019]
Cycling Embassy of Denmark—cycling-embassy.dk [Dec 2019]

Deutscher Tourismusverband—deutschertourismusverband.de [Sep 2019]
Die Bundesversammlung—Das Schweizer Parlament—parlament.ch [Sep 2019]
Direction Générale des Entreprises—entreprises.gouv.fr [Sep 2019]
Directorate-General for Passenger Transport, 1999 «The Dutch Bicycle Master Plan. Description and evaluation in an historical context»
Eidgenössisches Departement für Umwelt, Verkehr, Energie und Kommunikation—uvek.admin.ch [Sep 2019]
European Cyclists' Federation—ecf.com [Dec 2019]
Fahrradportal—Nationaler-radverkehrsplan.de [Sep 2019]
Federal Ministry for Economic Affairs and Energy, 2009, «Grundlagenuntersuchung Fahrradtourismus in Deutschland», Forschungsbericht no. 583
Federal Ministry of Transport, Building and Housing, 2002 «National Cycling Plan 2002–2012. Ride your bike! Measures to Promote Cycling in Germany»
Federal Ministry of Transport, Building and Urban Development, 2012 «National Cycling Plan 2020. Joining forces to evolve cycling»
Fédération Francaise de la Randonnée Pédestre—ffrandonnee.fr [Oct 2019]
Fietsersbond—fietsersbond.nl [Oct 2019]
Giacomel A. Moscarelli R (2018) «Ripensare la rete di infrastrutture turistiche eggere per rigenerare le aree interne. Il caso della dorsale cicloturistica VENTO» XXI Conferenza annuale SIU, Firenze—media.planum.bedita.net/7b/1b/Atti%20XXI%20Conferenza%20SIU%202018_Planum%20Publisher_W2.1.pdf [Gen 2020]
Groupe de travail pour le développement de l'usage du vélo, 2012 «Plan National Vélo»
Holland Cycling—holland-cycling.com/blog/274-lf-routes-to-be-overhauled [Oct 2019]
Landelijk Fietsplatform—fietsplatform.nl [Oct 2019]
Les Véloroutes et Voies Vertes de France—af3v.org/-Cahier-des-charges-national [Oct 2019]
Ministero delle Infrastrutture e dei Trasporti—mit.gov.it [Gen 2020]
Pileri P, Giacomel A, Giudici D (2015) «VENTO. La rivoluzione leggera a colpi di pedale e paesaggio» , Corraini editore
Radnetz Deutschland—radnetz-deutschland.de [Dec 2019]
Swiss Federal Council (2016) «Sustainable Development Strategy 2016–2019»
SwitzerlandMobility—Schweizmobil.ch [Dec 2019]
SwitzerlandMobility Foundation—Switzerlandmobility.org [Dec 2019]
Transportministeriet (2014) «Denmark—on your bike! The national bicycle strategy»
Vélo and Territoires—velo-territoires.org [Dec 2019]
Wandelnet—wandelnet.nl [Oct 2019]
Wikipedia D-Route—de.wikipedia.org/wiki/D-Route [Dec 2019]

The United Kingdom's National Cycle Network: Paths for Everyone, Past, Present and Future

Xavier Brice

Abstract The National Cycle Network (NCN) covers every region and nation of the United Kingdom. At its peak, it totalled over 16,500 miles, with routes on quiet roads, traffic-free paths and greenways. First conceived of in 1997, the network was initiated by the sustainable transport charity Sustrans; this organisation is still the custodian of the Network but it remains very much a joint production. Routes are provided and maintained by local government, private landowners large and small, volunteers and government and non-governmental bodies. It is a major collaborative achievement of civil society and in 2017 carried 785 million journeys a year, with an estimated annual contribution to the UK economy of £3.8 billion. This paper does three things: first recount the genesis of Sustrans and the NCN; second, explain the rapid growth of NCN in the early twenty-first century and its impact; third, explain the 2017–2018 review of the NCN and its re-visioning as a network of traffic-free paths for everyone—in many ways a return to its founding ideals and one that is as important as ever.

In January 2020 Sustrans employed over 600 people across the UK, with offices and teams in Scotand, Wales and across England including London and Bristol. Its work is made possible by over 3,500 volunteers and over 35,000 supporters. Sustrans has two priorities, creating liveable cities and towns for everyone, and creating paths for everyone—which is implementing the new vision for the National Cycle Network details of which are below. The charity's work is funded by a combination of grants from governments and charitable trusts, donations from individuals and contracts for advice and delivery of walking and cycling infrastructure from the government and other organisations. www.sustrans.org.uk has more information on the charity's strategy and current organisation.

X. Brice (✉)
Sustrans Limited, 2 Cathedral Square, College Green, Bristol, UK
e-mail: Xavier.Brice@Sustrans.org.uk

P. Pileri and R. Moscarelli (eds.), *Cycling & Walking for Regional Development*, Research for Development, https://doi.org/10.1007/978-3-030-44003-9_14

1 1976–1996: The Genesis of Sustrans and the National Cycle Network

As the United Kingdom (UK) approached the new millennium it had no coherent cycle network. There was patchy provision in a few towns and cities, notably the post-war new towns of Milton Keynes, Stevenage and Bracknell built with traffic-free networks, and campaigners in London had succeeded in getting some provision. But there was nothing nationally. In the 1930s, the government had proposed building segregated cycle lanes alongside the new trunk roads being rolled out to serve the growing number of motor vehicles. Some were built and sections still exist today, including alongside the A4 near London's Heathrow Airport; however, cycling organisations of the time saw these cycle lanes as a way to ban cyclists from the public highways. They fiercely opposed their construction and no coherent network emerged (see Reid 2015). This changed with the construction of the National Cycle Network, funded by a grant from the Millennium Commission in 1995, in response to a proposal from the sustainable transport charity, Sustrans.

Today, Sustrans is an established charity whose mission is to make it easy for people to walk and cycle. The organisation began in 1977, in the city of Bristol in the west of England. Set up by cycle activists and environmentalists in the wake of the oil crises, the organisation was initially called CycleBag and set out to 'demonstrate the benefits to health of bicycling and walking; the ecological and economic benefits of bicycling and walking; and the dependence on oil of our transport culture'. Under the charismatic leadership of one of the founders, John Grimshaw, the organisation developed rapidly. Grimshaw later recalled that he felt the campaigners needed a delivery arm, and in 1979 he set to work converting a disused railway line between the cities of Bristol and Bath. This was done in stages, with the help of local government, and built using largely volunteer labour, creating one of, if not the, first UK greenway converted from an old railway. It was no easy task. The first stretch was five miles long and it took persistence and imaginative thinking to convert the full 15-mile route. By 1986 it was complete, providing a traffic-free greenway for use by walkers, cyclists and wheelchair users between two cities, much to the bemusement of some local residents (see Fig. 1). The route now carries more people on bicycle, foot and wheelchair than the railway ever did.

By the time that the Bristol and Bath path was complete, work had started on a number of other paths and Cyclebag had turned into the charity Sustrans (a compression of 'Sustainable Transport'). In 1982 John Grimshaw had persuaded the UK Government's Department for Transport to commission a study on converting redundant railway lines into walking and cycling routes. The infamous Beeching report in 1963 had resulted in the closure of one-third of the country's railway network—a strategic mistake whose effects are still felt today. Other lines had closed prior to that; the railway mania of the nineteenth century had left a legacy of duplicative lines, including between Bristol and Bath. The prospect of lines reopening for trains

Fig. 1 Opening day of the Bristol to Bath railway path—6 August 1985. *Source* Sustrans archives

was remote at best in the early 1980s.[1] So these old routes were there for the taking, though doing anything with them would take considerable energy and commitment. The government decided not to adopt the findings of Grimshaw's study; in anticipation of this he had prepared feasibility studies for 30 potential routes, constituting an appendix larger than the report itself! Sustrans now got to work turning the appendix into reality. This was not a coherent route network. It was a series of short, opportunistic links driven largely by the availability of a redundant railway line, as can be seen from a very early scoping map from the Sustrans archives shown at Fig. 2. But it was a significant start.

One major coup was acquiring a redundant duplicative section of the East Coast Main Line between York and Selby in the county of Yorkshire. This was the very line that the famous Flying Scotsman train used to thunder along between London and Edinburgh, it connected two centres of population. Sustrans was established to enable this purchase from the government, and it was one of many land acquisitions that underpinned the early growth of the Network. Sustrans owns very little of the land that carry the traffic-free routes. Only 1.5% of the NCN today is owned by Sustrans. But the ability to purchase substantial sections of route unlocked larger schemes.

[1]Indeed, in 1982 the Serpell report to the UK Government proposed closing over 80% of the remaining route network—though fortunately this was not implemented. For more on this see Railways Archive (2019).

Fig. 2 Potential greenway routes on disused rail lines plotted in the 1980s. *Source* Sustrans archives

Small sections of land often had to be bought outright or licences negotiated with owners to ensure continuity; reluctant landowners were often more prepared to sell a section of land to a charity with no commercial interests, than to a company or government body.

Across the UK Sustrans was now building routes, guided by that 1982 study, and fuelled by volunteer labour and a wide range of funding sources. This was not a coherent network, and it was not backed by a single funder. But with communal energy and commitment routes were created. In 1983 a separate Scottish Railways study by Sustrans saw building commence in Scotland, with a path between Glasgow and Dumbarton, and in 1987 the Pilton Path in Edinburgh was built with the local cyclist activist group Spokes. Everywhere, local collaboration was key. Sustrans did not and could not operate alone. On the western edge of the Lake District, Sustrans worked with the charity Groundwork to build routes on railways leftover from long-closed Iron and Coal industries. And as these sections of greenway started to be joined up into longer route, Sustrans worked with and helped establish organisations specifically dedicated to certain routes—such as the Trans Pennine Trail which now runs from Southport on the west coast to Hornsea on the east, incorporating the York to Selby line and with numerous branches passing through the Pennine hill range, linking historic northern towns and cities with beautiful countryside. The Trail is managed by a partnership of 27 local authorities and supported by a very active Friends group that provides financial support for new projects and volunteers to help look after the routes.

These longer routes were no longer just greenways but started to combine them with routing along canal tow paths, and where there was no alternative—roads. By 1994 the first long-distance route was in place, the 132 mile long Coast to Coast or C2C, from Whitehaven in Cumbria on the west coast, to Tynemouth on the North Sea in the east. The C2C route also saw one of the earliest artworks commissioned by Sustrans, which was to develop into a programme across the Network called Art in the Travelling Landscape. Andy Galsworthy was commissioned to produce a site-specific sculpture at Leadgate in County Durham—see Fig. 3. These structures are still present today. They demonstrate the early ambition of Sustrans, and how these routes were not merely functional links, but rooted in and enhanced the history and culture of places that they connected.

In 1994 Sustrans submitted the idea for a National Cycle Network to the Millennium Commission. The Commission had been established by the UK Government to fund ambitious projects to celebrate the new millennium using proceeds from the newly created National Lottery. Examples of other funded schemes include the

Fig. 3 Andy Galsworthy sculptures at Leadgate on the C2C route. *Source* Sustrans archives

Eden Project in Cornwall and the Millennium Dome in Greenwich, London (now the O2 music venue and attraction). Sustrans's submission for the country's first National Cycle Network was certainly ambitious. The goal was to create a signed route network of 5,000 miles, half on traffic-free routes, the other half on quiet roads and all of it suitable for 'an unsupervised 12 year old'. The initial proposal covered Wales, Scotland and England but excited at the prospect of a truly UK-wide project, the Commission asked that Northern Ireland was included. As is so often the case, the proposal had been put in tight to the deadline and so it was hastily revised to include Northern Ireland and to this day, the NCN remains the only UK-wide physical transport network.[2]

There is no single inspiration for the NCN. Many European countries had national and regional networks by this point. Staff from Sustrans were travelling abroad to study best practice, and to help influence decision-makers in the UK Government—though John Grimshaw was careful which examples he used. Comparisons between the dearth of cycle infrastructure in the UK and the high-quality networks in the Netherlands would draw the response that Britain was not flat. So, instead, he would point to the Swiss networks. But it was the Danish cycle network that inspired the numbering and signage of the UK's Network—red numbers on blue signs.

The proposal was accompanied by a trail-blazing ride across the length of the country, from Inverness in the Highlands of Scotland, to Dover on the Kent coast. These were not protest rides but intended to demonstrate the level of popular support, and engage local politicians and other influencers. Of course they also acted to highlight the lack of infrastructure and inadequacy of existing conditions, and by going down sections of land that Sustrans aspired to convert into NCN gave a glimpse of what could be possible. The Commission was won over and in 1995 Sustrans was awarded a large cheque. The Network was born.

2 1996–2016: Growth and Impact

One critical question for Sustrans in 1995 had been how much money to ask for. The submission was grounded in knowledge gleaned through route-building to date, and estimates and assumptions. John Grimshaw settled on the figure of £42 m. Why? Not because it would take £42 million (m) rather than £43 m or £41 m but because that was the answer to the ultimate question of life, the universe and everything in Douglas Adam's cult radio show and novel, The Hitchhiker's Guide to the Galaxy. Given that large infrastructure projects' final costs bear little resemblance to their original estimates, some would argue that this is as sound a logic as any. Many more would agree when comparing the original 5,000-mile long proposal with the NCN's subsequent growth and impact.

---

[2]The road network is administered separately across Northern Ireland, Scotland, Wales and England, as is the canal network, whilst the rail network is under separate governance in Northern Ireland.

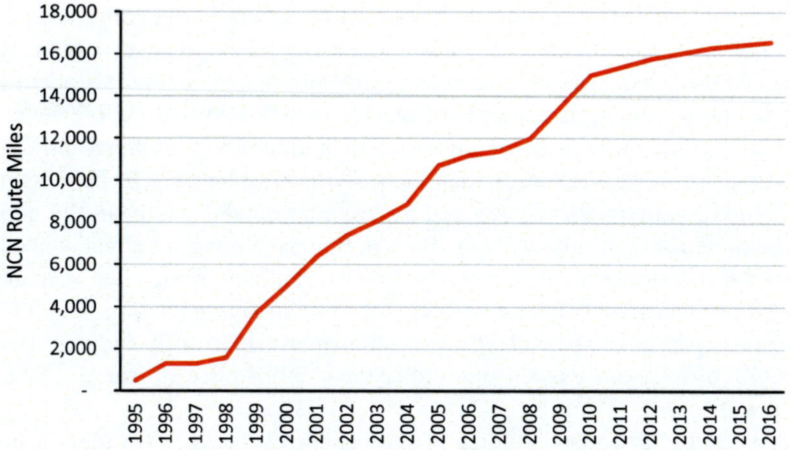

Fig. 4 Growth in National Cycle Network route miles 1995–2016. *Source* Sustrans data

By the end of the year 2000, the NCN had reached 5,000 miles. The target had been delivered and over the next eight years the Network grew to reach over 15,000 miles—see Fig. 4. This consisted of 10,000 miles of national route, and 5,000 miles of secondary, regional routes. The national routes being the core trunk routes, with single or double-digit numbering. The secondary routes were intended to provide regional connections, and typically had triple-digit numbering. Later on the distinction between regional and national routes disappeared and, collectively, these 15,000 miles of Network grew to a total distance of over 16,500 miles, to a point where half of the population of the UK lives within one mile of the NCN.

This growth was made possible through cooperation with a range of organisations, and securing funding far in excess of the original Millennium Commission grant. Money tends to beget money, and being in a position to develop routes on other bodies' lands meant that Sustrans was often able to attract matched funding, either from that body or from other grant-making organisations. Individual philanthropy had always played an important role, with a supporters programme launched in 1992 that provided a way to connect anyone who wanted to financially contribute to the Network, and also providing a core network of support out amongst the general public. The coming together of individual support, grants and match funding from other bodies is exemplified by Sustrans' successful campaign in 2007 to win the People's £50 m Lottery Contest run by the Big Lottery (the successor to the Millennium Commission, and now called Community Lottery) in 2007. Once shortlisted entrants had to win a popular vote. Sustrans' bid was to extend the NCN providing more local links to connect places and communities under the title of Connect2. Local communities were engaged in the build-up to the vote, and an additional £100 m of local authority funding was identified, with £126 m eventually secured. That the Connect2 bid won also shows the popular appeal of walking and cycling routes and the NCN. This was not, and is not, an enthusiast's network. It was, from the outset, by the people and for the people and remains so today.

Connect2 created 84 new walking and cycling networks, integrated with the NCN, 80 major new safe crossings of roads and built or renovated over 100 bridges. The impact of these local connections was significant. A five-year independent research programme by the iConnect consortium (2016) showed that people living within 1 km of the new infrastructure increased their time spent walking and cycling by an average of 45 min per week than those living 4 km away. The benefits were not distributed equally but were enjoyed by less represented groups in the population. For example 100% of survey respondents in Belfast, Northern Ireland, who classified themselves as 'looking after home or family' said new routes had increased their regular physical activity levels, with 90% of women saying the same vs 80% of men. Whilst completion of a historically long unfinished bridge in Glasgow created a route that 78% of respondents said was either their only source of exercise or a wholly additional source.

There have been many studies of the benefit of the 16,500 miles of the NCN. The most recent research by Sustrans in 2016 calculated that the annual economic impact of the Network was £3.8bn. This breaks down into £1.3bn from economic savings through the trips made, driven largely through health benefits, but also reductions in congestion, CO2, air quality and increased journey quality. The remaining £2.5bn is direct economic contribution from leisure and tourist trips, and considers accommodation, transportation, food and drink costs. Whether indirect or direct, these impacts are felt locally. Unlike major road or rail infrastructure projects that require massive investment to benefit a particular corridor or location, investment in a cycle network spreads the benefit locally, and access is local. No need for a motorway junction or station. The same is true of the tourism impact. In 2012 a study for the European Parliament calculated that cycle tourism was worth €44bn annually to the European economy. Not only is that more than the cruise industry, but it is a far more sustainable and equitable impact. The shocking environmental impact of the cruise industry is well known but the downside of the concentrated economic impact on certain locations is now a major issue.[3] If tourist spend is like blood for the geo-economic body, then the cruise ships leave huge bruises. Indigestible and painful concentrations of spend with a bruising impact. Cycle tourism, on the other hand, spreads that blood across the body. Cycle networks act as arteries and thin capillaries distributing money into regions and communities that would be passed by most other forms of tourism investment. The NCN is a founding part of the Eurovelo network which helps attract foreign users and visitors. In 2015 Sustrans estimated the indicative value of leisure cycling and cycle tourism on the 2,371 miles of NCN in Scotland alone was £345 m, with £116 m of this coming from tourists and the remaining £229 million from home-based visitors.

Such economic analysis is one way to capture the broad range of the Network's impact but what it cannot convey is the huge impact that the Network has had on the development of cycling infrastructure and cycling culture in the UK. When the

[3]For example, in 2017 Carnival Corporation, the world's largest luxury cruise operator, emitted nearly 10 times more sulphur oxide around European Coasts than all of the cars in Europe. See Transport and Environment (2019).

Network started there was little in the way of dedicated infrastructure. One of the goals of the NCN was to act as a catalyst for local networks. The world-class cycle infrastructure in London built over the last decade has its roots in the NCN, as this led directly to the concept of a London Cycling Network—or LCN. Although the LCN was never finished, replaced instead by higher quality Cycle Superhighways and Quietway routes, it provided the foundation for those routes in many places, and critically helped generate further demand for cycling infrastructure in the UK's capital city. London is an instructive example of both the trail-blazing impact of the NCN but of how that very role has left the Network wanting in many places.

With no grants to compare to the Connect2 award in 2007, and with few levers of control over the Network—the vast majority of which runs on other bodies' roads and land—the charity had not been able to keep pace with the quality of infrastructure seen in, for example, London or Cambridge. It must be emphasised that these cities are isolated examples in a country where provision is still woeful compared to most of Northern Europe but they also highlight a key issue with the NCN. In hindsight, the Network's rapid growth in the early 2000s, meant that quality was too often compromised. The original plan for a 5,000-mile network, half traffic-free, half on-road evolved into a network three times as long, but with only twice as much traffic-free route. And almost two decades of increases in road traffic and congestion across the UK means that many of the roads that were calm and quiet in 2005 were now anything but. For all the Network's pioneering impact there was a growing need to work out how to develop the NCN for the future and ensure it's continued impact.

3 2016 Onwards: Review and Re-Visioning as Paths for Everyone

In 2015 a thorough audit of the NCN was started by Sustrans to record the state of the network. A combination of qualitative and quantitative criteria fed into seven levels of service measures: Accessibility (can all users access and travel along it); Surface quality; Traffic-related safety; Way-finding and signage; Flow (is the path wide enough to enable all users to travel in comfort and are there any barriers impeding continual movement); Social Safety (whether it feels a safe place to be); and Place (is the route attractive and interesting). The audit revealed that although much of the one-third of the Network that was traffic-free was in good condition, most of the remaining two-thirds of on-road Network was in very poor condition (see Fig. 5) Furthermore, there were over 16,000 physical barriers on the network. Often put up to prevent illegitimate access, typically by off-road motorbikes, these barriers impeded progress, especially for those users on adapted cycles, with pushchairs or in wheelchairs and mobility scooters.

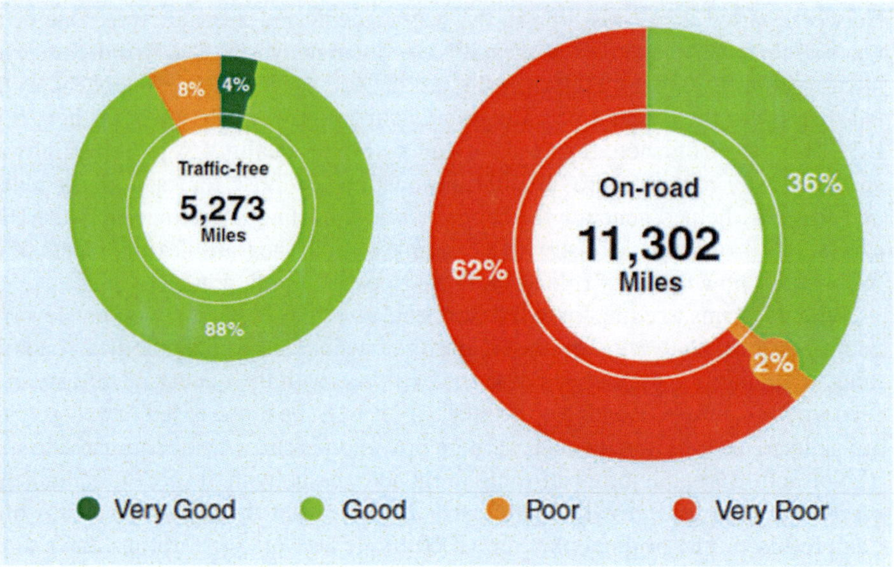

Fig. 5 Future route type on the National Cycle Network—current and target. *Source* Sustrans, Paths for Everyone, Sustrans 2018

Despite this, usage figures for the Network were high.[4] In 2017 it carried an estimated 4.4 million users, on 786 million journeys—for comparison, the entire British rail network carried 205 million passenger journeys in 2016/17.[5] And people knew about it, with 1 in 4 adults having heard of the NCN when prompted by name. Significantly, more than half of the trips made on the Network were actually done on foot—with walking making up 55% of trips in total, compared to just under 45% by cycle.

The audit and survey data fed into a comprehensive review of the NCN. This was led and largely funded by Sustrans but it was supported by funding and participation from the UK, Scottish, Northern Ireland and Welsh Governments. Crucially, it also involved a range of organisations drawn from major landowners who had NCN

[4]The usage estimates for the National Cycle Network are calculated from a network of counters distributed across the Network, covering a representative sample of location and route type—for example, rural greenway, or urban on-road with pavement. These are then scaled up and adjusted to provide an estimate for the entire Network. These are supplemented by route-intercept surveys of individual users which also provide demographic and other quantitative date on the people who use the NCN. Full primary monitoring across the 16,500 miles would require significant investment, and the rapidly developing field of crowdsourcing such data through mobile phone based applications or other GPS user-devices will capture only particular segments of users; though Sustrans and its partners continues to investigate how to improve data collection on the Network. A journey is any trip by an individual, for any purpose, whether utility—such as commuting or a journey to school, or leisure—walking the dog or a cycle-tourist journey. Trip and journey are used interchangeably here as we do not distinguish between journey legs of a longer trip.

[5]Rail passenger journey data from the Office of Rail and Road (2020).

running on their land or roads, including local authorities, the National Trust, and the Canal and Rivers Trusts, to user groups. The conclusions were therefore not Sustrans' but those of the users and partner organisations who had helped build, grow and nurture the Network. The conclusion is encapsulated in the new vision for the NCN published in the review report, called Paths for Everyone (Sustrans 2018, p. 10):

> A UK-wide network of traffic-free paths for everyone, connecting cities, towns and countryside, loved by the communities they serve

This vision does not specify what type of trips the network aims to serve. This is a deliberate choice. Sustrans thought long and hard—together with the Network's partners—on whether we should define a key purpose by type of journey. The decision was taken not to. The national road and rail network carries people making trips for all sorts of purposes. The NCN is arguably a more diverse network than either, with a multiplicity of route types and working at both a very local level, and at the supranational level. Whilst it is not a substitute for fine-grain town or city-based networks (on which the UK lags behind the rest of northern Europe and which are absolutely critical) it is local. It provides linking routes within and between communities. It also serves to connect counties, regions and indeed nations. So the Network is used for cycle tourism, for commuting, for school trips, by dog walkers and by people just out for a stroll or cycle. What is more important is the quality of the experience, the ethos, and the involvement of community. And these three things sit at the heart of the new vision for the NCN.

The ethos of 'for everyone' is a change and is encapsulated in the cover of the report—see Fig. 6. This is not for lycra-clad racing cyclists. The design speed is 15 km/h, which is relatively low in cycling terms. It is for everyone, regardless of age, experience or ability, to use on foot, on a bicycle, wheelchair, scooter or where possible, by horse. This means removing or redesigning the 16,000 + barriers that the audit mapped, and it means working hard to broaden the Network's appeal. Community has been at the heart of the inception and growth of the Network. Since 1998 Sustrans has had a Ranger's programme, where volunteers act as NCN Rangers, responsible for monitoring the upkeep of sections of the Network. Volunteering at Sustrans has since grown to encompass far more than the NCN, but empowering Rangers to work with other volunteers (from Sustrans and partner organisations) and engage their local communities is at the heart of the new vision for the Network, and one example of how it goes back to the roots of the Network.

Perhaps the most notable aspect of the new vision is that, for the first time, this puts a traffic-free ethos at the heart of the whole Network—and as outlined in the genesis of the NCN, it was the development of traffic-free routes that led to the inception of the NCN and provides its backbone. The traffic-free routes carry some of the highest volumes of people on the NCN, and it is these sections that make the Network special. A survey of over 6,000 users and potential users revealed that 81% of people wanted to see 'more traffic-free routes where everyone feels safe to get around'. There is overwhelming evidence that fear of road danger puts people off cycling and that what people want to make more trips under their own power,

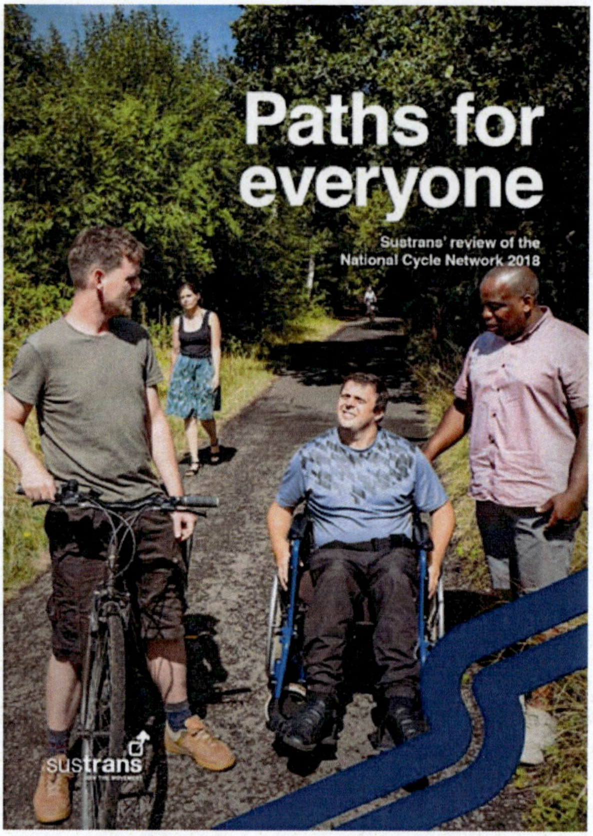

Fig. 6 The image chosen for the cover of the Paths for Everyone report on the review of the Network (Sustrans 2018) encapsulates the traffic-free vision of a network for all

whether on foot or by bicycle, is routes away from the noise, pollution and danger of motor vehicles. It is also on the traffic-free parts of the NCN where a deeply rooted sense of place is most important and most fully realised. People sharing a path typically need to interact in a way that they don't on the road. They share a greeting, or ring a bell or say excuse me. At a time of growing division, and where our lives can be stratified and segmented, traffic-free paths can be places where we can live well together by moving well together. It was through traffic-free routes that the Network began and it now needs to return to that. Sustrans is not aiming to make all the Network traffic-free. It has the objective of delivering a network that is two-thirds traffic-free by 2040, with the remaining third on quietways (see Fig. 7).

Quietways are defined as roads with a speed limit of 40 miles per hour (mph) and a daily traffic flow of less than 1,000 vehicles in rural areas, and 20mph and 2,500 vehicles in built-up areas, together with appropriate signage, width and visibility. In short, safe enough for a 12 year old to use independently—a return to the founding vision—and where you would feel comfortable pushing a small child in a pushchair,

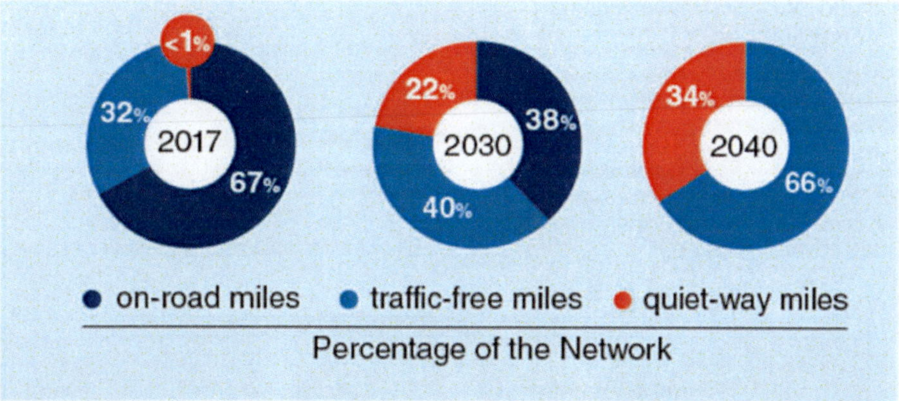

Fig. 7 Condition of the National Cycle Network by type of route in 2017. *Source* Sustrans, Paths for Everyone (2018)

or using a wheelchair. As well as route type, the Review specified quality, with nine design principles as seen in Fig. 8.

Improving route type and quality were a key focus of the review but by no means all of it. In conclusion, the review made 15 recommendations which Sustrans and its many partners are now in the process of implementing.[6] And echoing Grimshaw's original report on converting railway routes in 1982, the NCN review included appendices outlining 55 activation projects across the UK that looked to address the issues identified in the report and the desired changes—including converting on-road to traffic-free, tackling busy junctions, removing barriers and improving the quality of poor traffic-free sections. The majority of these projects were unfunded at the time of the report's publication at the end of 2018 but by the end of 2019 nearly all of them had funding lined-up (largely but not only from the Department for Transport, Transport for Scotland and local authorities) and many were well underway. This

[6]The 15 recommendations are detailed on page 7 of the Paths for Everyone report (Sustrans 2018): 1. Set the Tone for Harmonious use of the Network; 2. Remove or redesign all 16,000 barriers on the Network to make it accessible to everyone, with no barriers in place for continuous travel; 3. Transform the Network by replacing existing on-road sections with new traffic-free paths or by creating quietway sections so it is safer for everyone; 4. Ensure that where the Network is on a quietway section the speed limit is 20mph in built-up areas and 40mph in rural areas; 5. Improve safety at crossing where the Network crosses road or railways; 6. Adopt a new quality standard to ensure paths widths and surfaces are built for everyone; 7. Improve signage so everyone can use the paths without a map or smartphone; 8. Deliver over 50 activation projects across the UK by 2023 to improve the Network and demonstrate change; 9. Introduce a process for de-designation of parts of the Network that cannot be improved—and a clear process for incorporating new routes that fill gaps or make new connections; 10. Make it easier for people using the Network to feed back on its condition—and use this insight to improve it; 11. Promote the Network to new users; 12. Encourage greater community involvement in designing, developing and maintaining the Network; 13. Provide open data on the Network; 14. Report regularly on the impact of the Network in improving everyone's lives and places; 15. Establish clear governance to bring together land managers, funders, users and others to deliver these recommendations in partnership.

Fig. 8 A quality standard for the National Cycle Network. *Source* Sustrans, Paths for Everyone, 2018

was not a review to file away, but a new vision, the start of a new plan to overhaul the National Cycle Network and create Paths for Everyone.

Delivering the Paths for Everyone's vision and the recommendations of the report will require partnership. The Network came about through individual vision and gumption. The need for clarity of purpose and the personal leadership to push through a big idea cannot be underestimated. But that is not nearly enough. The Network was made possible by collaborating with a range of different governmental and non-governmental agencies and partners. Such an endeavour is too big for one organisation or even for one sector. It takes nations, regions, and individuals in thousands of communities to create a lasting network that is of such value to so many people.

References

Brice X (2018) The UK National Cycle Network. Paper presented at the 29th International Cycle History Conference, Guildhall, London

European Parliament (2012) The European Cycle Route Network Eurovelo Study. https://ecf.com/sites/ecf.com/files/EP%20study%20on%20EuroVelo%20network.pdf. Accessed 10 January 2020

iConnect Consortium (2016) Fit for Life: Independent research into the public health benefits of new walking and cycling routes. https://www.sustrans.org.uk/media/3691/sustrans-fit-for-life.pdf. Accessed 13 December 2019

Office of Rail and Road (2020) Passenger Rail Usage. https://dataportal.orr.gov.uk/statistics/usage/passenger-rail-usage/. Accessed 24 January 2020

Railways Archive (2019) Available via http://www.railwaysarchive.co.uk/docsummary.php?docID=21. Accessed 13 December 2019

Reid C (2015) Road were not built for cars: how cyclists were the first to push for good roads and became the pioneers of motoring. Island Press, Washington DC

Sustrans (2018) Paths for Everyone: Sustrans' review of the National Cycle Network 2018. https://www.sustrans.org.uk/media/2804/paths_for_everyone_ncn_review_report_2018.pdf. Accessed 19 January 2020

Transport and Environment (2019) One Corporation to Pollute Them All: Luxury cruise air emissions in Europe. https://www.transportenvironment.org/publications/one-corporation-pollute-them-all. Accessed 10 January 2020

Trans Pennine Trail (2020) https://www.transpenninetrail.org.uk/. Accessed 2 January 2020

Zovko I (2013) The value of cycle tourism: opportunities for the Scottish economy. http://transformscotland.org.uk/wp/wp-content/uploads/2014/12/The-Value-of-Cycle-Tourism-full-report.pdf. Accessed 10 January 2020

The Singularity of the Camino de Santiago as a Contemporary Tourism Case

Rubén C. Lois González and Lucrezia Lopez

Abstract The present work analyses the uniqueness of the Camino de Santiago (or The Way of Saint James), a religious and medieval pilgrimage route that has become a contemporary tourist product with a deep power of attraction. To this end, a series of factors that have contributed to the change and renewal of its nature are detailed. Firstly, we explain the public promotion of the Camino as a catalyst for a process of physical construction of a pilgrimage route. Secondly, we focus on studying the Santiago pilgrim's motivation, as it is also a relevant factor in determining slow mobility in urban and rural areas. Finally, these factors interact with each other and result in the definition of the pilgrim as a new tourist who walks or rides a bicycle. All of the above leads to consider new social and territorial dynamics.

1 Introduction: The Historical Construction of the Camino

From a historical point of view, the city of Santiago de Compostela, and its cathedral, was the final destination of the main European pilgrimage route in the Middle Ages, and more so since the twelfth century (Moralejo 1993; Soria y Puig 1993). This fact, demonstrated by many investigations, served as a basis for the real (re)construction of a route of medieval and religious origin in the second half of the twentieth century and in particular since 1990, that has turned it into a cultural and tourist itinerary, in accordance with the demands of the new contemporary (post-secular) *pilgrim* (Lois 2013; Lois and Lopez 2012). This new route has been designed and created by the public authorities, first to extol its historical and heritage value, and also to favour slow mobility, which is one of the main attractions and keys to the success of the Camino de Santiago today (Lois and Somoza 2003; Lois et al. 2016; Somoza and Lois 2018) (Fig. 1).

R. C. Lois González · L. Lopez (✉)
Department of Geography, University of Santiago de Compostela, A Coruña, Spain
e-mail: lucrezia.lopez@usc.es

R. C. Lois González
e-mail: rubencamilo.lois@usc.es

P. Pileri and R. Moscarelli (eds.), *Cycling & Walking for Regional Development*, Research for Development, https://doi.org/10.1007/978-3-030-44003-9_15

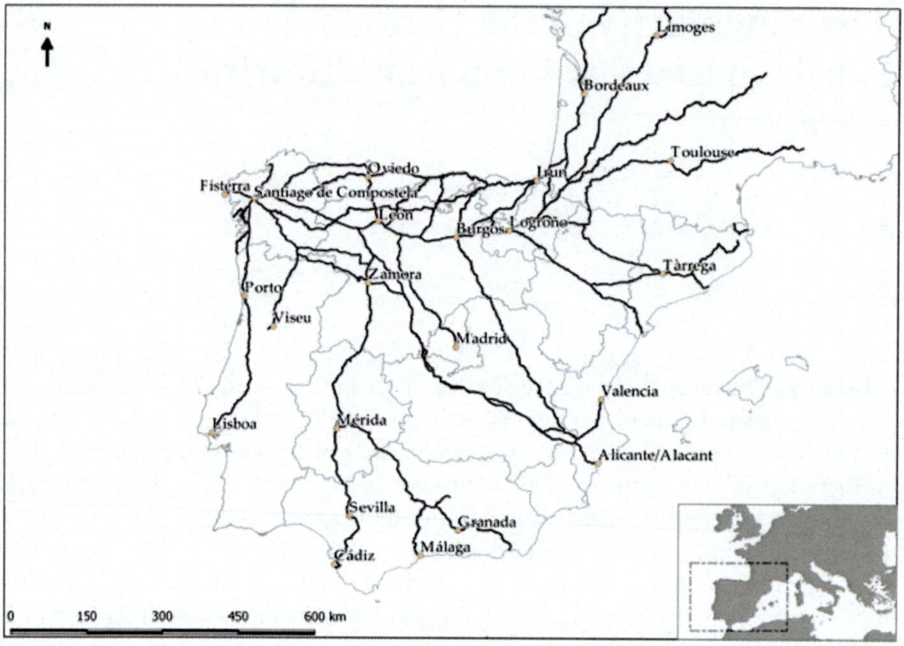

Fig. 1 The Ways to Santiago de Compostela (*Source* author's own work)

Although the myth of the discovery of the remains of the Apostle James the Great dates back to the beginning of the ninth century, it would not be until the twelfth century when after intense negotiations and the payment of significant sums of money, the Papacy of Rome recognised the authenticity of the relics in Santiago de Compostela (Falque 1994; Moralejo 1993; Rey Castelao 2006). From that moment on, religious trips multiplied throughout Europe and traditional pilgrimage routes to Santiago were established. Among them, the French Camino,[1] starting from Paris, Vezelay, Arles or Le Puy and delimited and promoted by Pope Calixto III in book V of the *Codex Calixtinus* as the ordinary Camino to Santiago de Compostela, became the most renowned. In present-day Spain, the route crosses the Pyrenees, the high valley of the Ebro River, the Plateau and enters Galicia through the Cebreiro Mount, to end in a fully Atlantic landscape, green and full of hills and topographic undulations (Somoza and Lois 2018; Soria y Puig 1993; Torres et al. 1993). This pilgrimage route was immediately protected by ecclesiastical and civil authorities, with new cities and towns alongside, in addition to monuments and road engineering works, which reinforced its heritage value. Today, the nuclei of the Camino are magnificent examples of well-preserved historical centres, Romanesque churches, Gothic cathedrals and unique buildings and Baroque monasteries and large civil constructions, alongside

[1]It is the most famous of all the routes leading to Santiago. The majority of pilgrims choose this itinerary, which usually begins at Saint Jean Pied de Port, in France.

bridges, fountains and historic lodges in a space of massive heritage value (Franco and Tarrio 2000; Somoza and Lois 2018; Soria y Puig 1993).

Considering the enormous weight of history in shaping the Camino, there is a broad academic consensus that the popularity of the route has been (re)invented in contemporary times (Lois 2013; Lois and Lopez 2012, 2019). It all began with the affirmation of popular religiosity of the late nineteenth century and, in the case of the Camino, its first key period was during the Franco regime. The authoritarian regime after the Civil War in 1939 used the myth of Santiago (James the Apostle) as a powerful war image against communists and freemasons. On a practical level, the government continuously invested significant sums of public money in the historicist (re)monumentalisation of the Camino in a vast work directed for many years by the architect F. Pons Sorolla (Castro 2010, 2013). Interestingly, the democratic transition and the incorporation into European institutions reaffirmed this idea of assigning large sums of public money to the construction of a pilgrimage itinerary (now defined as tolerant, multicultural and multireligious, under the European Cultural Itinerary certification of the Council of Europe), but focusing more on the material aspects of the secular route. Regional governments and, to a lesser extent, national governments, began to neaten the path of the Camino, building or promoting shelters in its sections, and signalling a route that was meant to be the meeting place of thousands of contemporary travellers from all over the world, united by their interest in walking and the contemplation of the landscape (Bermejo López 2001; López Trigal 1993; Pichel 2004; Rodríguez 2004).

2 The Public Promotion of the Camino: The Physical Construction Process of a Pilgrimage Route

As mentioned above, the (re)invention of the Camino de Santiago is barely a century and a half old and has focused on a succession of actions developed since the mid-twentieth century. Thus, together with the initial promotion of parish and religious pilgrimages of the late nineteenth century, at the end of the Civil War the political use of the myth of Santiago was useful to the Franco dictatorship. General Franco intended to justify the origin of the new regime (a cruel Civil War) on the metaphor of a crusade, where national and Catholic forces symbolically supported by Santiago the warrior had triumphed. On the material level, this powerful discourse preached from power explains why some distinguished Catholic professors of the time, and as soon as the 1940s, began the process of historiographic and physical recovery of the route, which materialised in a monumental work (Vázquez de Parga et al. 1948-49). The French Camino was redefined, recovering the layout of the *Codex Calixtinus*, and from that moment the government granted broad powers to the architect Francisco Pons Sorolla to undertake a monumentalising and historicist intervention in the cities and towns of the Camino (Castro 2010, 2013). Among other actions, the historic centre of Santiago was ordered, the traditional village of O Cebreiro was

Fig. 2 Main square of Portomarín (*Source* author's own work)

moved and rebuilt[2] and the new layout of Portomarín was designed[3] thanks to great sums of public funds aimed at recovering the medieval and Christian essence of the Camino (Fig. 2), through the unity of style or the exaggeration of the historicity of its elements (Castro 2010; García Cuetos et al. 2010). This complete reconstruction programme, reinforced by a surge in Jacobean research (in a period of markedly historical and philological component), resulted in the emergence of the first contemporary pilgrims. They were either students of the history of the Camino, people of entrenched religious beliefs, or both. As a result, both the recovery of the walking route to Santiago and the first associations of friends of the Camino appeared between the late 1950s and the 1960s (Lois 2013; Lois and Santos 2015).

From these foundations, the number of walkers to Santiago continued to grow while the route was being developed in some sections, protecting the pilgrims from road traffic, and the process of monumentalisation and patrimonialisation was taking place. Travellers from different European nations and Spanish regions were increasing in numbers, and gave the pilgrimage a much more tolerant, secular component, ultimately a contemporary significance to the route. This new reality was used by the public institutions of the Spanish transition and the following democratic period to redefine the Camino de Santiago, as the historical axis that facilitated the construction of Europe and the integration of the Iberian Peninsula into Europe

[2]It is the gateway to the French Camino in Galicia, located in the province of Lugo.

[3]Portomarín is a historical complex along the French Camino, in the province of Lugo. In 1963 a new Portomarín was completed and inaugurated, formed by several historical-artistic buildings of the old town, threatened by the waters of the Miño River.

(Caucci 1993; Lois 2013). Although the Holy Years[4] of 1976 and 1982 were not remarkable, they were the prelude of the recognition of Santiago's historic city as a World Heritage Site in 1985 and the consideration of the Camino as the main European Cultural Itinerary by the Council of Europe in 1987 (Martí 1995; Valle 1985). Between 1985 and 1987, Spain joined the European Community and the Camino was the theme chosen by the Spanish Government for a great exhibition in Brussels, which commemorated this historical event. Undoubtedly in the 1980s, the municipal government of Santiago, the newly founded Autonomous Region of Galicia and the national government of Madrid, began to collaborate and agree upon actions for the improvement and promotion of the Camino, in a practice that remains today. The number of pilgrims continued to rise to more than 10,000 in the early 1990s. But the great change came with Holy Year of 1993, known as Xacobeo 1993.

The celebration in 1993 of a Holy Year, after eleven years of no celebration of the day of Santiago landing on a Sunday, was used by the regional government of Galicia to launch a powerful campaign for tourism promotion and systematic intervention on the Camino, which was promoted under the Xacobeo name. This denomination used the Galician language (supporting the identity of Galicia), resorted to the qualifier derived from *iago* or *iacobo* (away from the most Catholic name of Santiago) and created a powerful advertising brand linked to the date of celebration, the Holy Year 93 (Xunta de Galicia 1993, 1994) (Figs. 3 and 4). For its design, a consolidated and conservative Galician government was inspired by the celebrations of the previous year (the Olympic Games of Barcelona and the Universal Exhibition of Seville) and continued to defend the institutional consensus (with a boost to the Royal Patronage of Santiago and the creation of the Santiago Consortium, to act in the old city centre) (Real Decreto - Royal Decree 260 1991; Xunta de Galicia 1993). The bulk of the funding was allocated to restoring the route: adapting the paving, improving traditional facilities (fountains, bridges, etc.), restoring churches, creating a network of public hostels for pilgrims that encouraged the journey on foot or by bicycle with guaranteed overnight stays, etc. The Camino was prepared as a great stage to march on foot and all this work was accompanied by an impressive advertising campaign that served to: present Galicia as a settled regional and historical reality, boost tourism in the region as a green destination and full of identity attributes and favour the great increase in the number of pilgrims (which went from more than 10,000 to never lower than 100,000) (Lois et al. 2018).

The success of *Xacobeo 93* was such in Spain and Portugal (there is more scepticism regarding its international impact), that suddenly the Camino de Santiago was placed as one of the main, and novel, tourist products of the country, when Public administrations were focusing on reducing dependence on Sun and Beach destinations in destination planning (Consejo Español de Turismo 2007; UNWTO 2005). The impressive impact justified that during the turn of the century, the Holy Years of 1999 and 2004 were understood as a continuation (more festive and expensive) of *Xacobeo 93*. The popularisation of the Camino in Spain and Portugal has been complemented by a cultured projection, and less subject to large advertising

[4]Jacobeo Holy Year is one in which July 25, the day of Santiago, falls on Sunday.

Fig. 3 Xacobeo 1993 logo (*Source Xacopedia*)

campaigns, of the route in the international context. Here, the associations of friends of the Camino, university courses, books and bestsellers, movies and blogs, have encouraged the pilgrimage of the Camino, from a more plural discourse, decentralised and insistently resorting to new media (Xunta de Galicia 1993, 1994). The number of users of the route to Santiago has not stopped increasing, with a maximum in the Holy Year of 2010, which has already been surpassed in the last campaigns (Oficina del Peregrino, Pilgrim Office 1990–2019). This fact generates an obvious problem that has become one of the main challenges of the present public policies: the Camino, and especially the historic centre of Santiago, are suffering from an overcrowding that can ruin the contemporary experience of the post-secular pilgrimage (Lopez et al. 2019). For this reason, the *production* of new routes has multiplied in recent times, some of them with a historical tradition of pilgrimage (the Northern Camino, the Primitive from Oviedo, the English from the ports of A Coruña or Ferrol, etc.) or that look for contemporary landscape and travel experiences, such as the Camino to Finisterre and the Portuguese, both of which are incredibly popular. In this Camino, Galicia has become a scenario for walking or cycling, with roughly ten routes, which already cross almost 40% of the municipalities in the region.

Fig. 4 El Pelegrín (The Pilgrim). Official mascot of *Xacobeo 93* (*Source Xacopedia*)

3 The Motivation of the Pilgrim to Santiago: Slow Mobility in Urban and Rural Areas

The previous pages have highlighted *the production process* of the Camino de Santiago: how the route has been ordered, how it has been financed, and how the areas have been monumentalised, promoted, new routes have been recreated, etc. Together with this *production*, it is now essential to analyse the keys to the success of the pilgrimage to Santiago, an issue that has already been addressed in previous works (Lois 2013; Lois et al. 2016; Lois and Lopez 2012). In these works we have been able to conclude that five main attributes explain the current success of the Camino de Santiago, based on two classic anthropological concepts, enunciated by V. Turner and recreated by other pilgrimage specialists in more recent times (Coleman and Eade 2004; Turner 1974). The first of these is that of liminality, which interprets life as a path through which a person passes through and where there are moments, rites of passage according to the author, which mean the transit between two distinct stages of existence (Lois and Lopez 2019; Turner 1974). In this way, both the vital meaning of the Camino, as a direction towards the following, and as an experience that comes to forge the personality in the transition from youth to adulthood, when ones working life ends and retirement begins, or after a deep crisis (a divorce, overcoming a serious illness, etc.), are understood. Numerous in-depth studies of the profile of the Santiago traveller show an over-representation of young people in the completion of studies, older pilgrims or people looking to open a new page in their lives (Lois et al. 2016; Lopez et al. 2017; Observatorio do Camiño 2007–2010). The second great concept contributed by Turner and anthropology is the sense of *communitas*, which permeates the contemporary pilgrimage. On the Camino, people meet, make friends and share reflections and experiences, especially among walkers, who encounter each other during the walk. At the same time, pilgrims speak, exchange opinions or try to meet local people along the route. Thus, a *communitas* is formed between the pilgrims (who will end up getting to know each other well, sharing blogs or simply memories), and between the walkers and the locals (who run bars or shops where they return to, send letters and photographs, etc.). In our opinion, the theoretical framework formulated by V. Turner clarifies the existential, curative and experiential content of the walk to Santiago. It explains the intimate motivations of its success (and overcrowding), and coincides with other more recent interpretations of the current tourist experience (Urry and Larsen 2011) and the contemporary sense of our relationship with places (Creswell 2004).

From the concepts of liminality and *communitas* it is possible to enumerate the five fundamental attributes in the contemporary revaluation of everything related to the Camino de Santiago. Undoubtedly the first, slow mobility, links with the need to break with the daily pace of life and take time for oneself, characteristic of contemporary tourism. People today usually live in cities, with scheduled timetables and move continuously from one place to another using mechanical means of transport. The walk, the pilgrimage to Santiago, entails a radical break with this everyday life. It implies mental rest, continuous physical exercise that takes us to our corporeality, returning to the human scale of things and the sensations, moving at 5 or 6 km/h, with enough time to think about the surroundings and the nature we are part of. To some extent, the slow mobility of the Camino links with the philosophical concept of *flâneur*. The men and women who have decided to travel the route experience a return to their own material essence, a direct and peaceful relationship with the places they cross, and the real dimensions of time and space. These facts make the Camino, the post-secular pilgrimage, deeply attractive, and explain its current success in urbanised and technified societies, as numerous studies on the subject have concluded (Badone and Roseman 2004; Coleman and Eade 2004; Greenia 2014).

A second attraction of the route is related to the possibility of enjoying one's surroundings and the contemplation, particularly when travelling on foot. The rural spaces, small forests and villages gain attractiveness and magnitude depending on their landscape value. It is significant to see how a non-existent attraction in medieval pilgrimages, the landscape that was not even referred to in the *Codex Calixtinus*, has become an important motivational element of the walk to Santiago in the present. In this sense, men and women of the twenty-first century, urban, with jobs and subject to daily routines, are eager to enjoy the outdoor spaces full of native flora and fauna, the historical towns and cities, and a reunion of what is perceived as natural in a Camino that crosses mountains (the Pyrenees and the entrance to Galicia), flat areas with ochre colour tones (the Castilian Plateau), river valleys (in La Rioja and Navarra) and Atlantic settings (in Galicia) (Badone and Roseman 2004; Greenia 2014; Lois et al. 2016).

Undoubtedly, slow mobility and contemplation of the landscape, together with the feeling of the body marching for hours, supports the idea of living a spiritual experience (Lopez et al. 2017). The Camino provides many hours to think while walking, also the landscape colours and tones stimulate the sensations, which together with people's beliefs (in general not particularly religious, but with a certain sense of the transcendent), explain that among the motivations of the current pilgrimage the spiritual one stands out (Observatorio do Camiño 2007-2010). A polysemic spirituality, since it implies both a weak permanence of the religious and a return (pantheist?) to nature and the possibility of thinking, of rejoining one's own values, problems and beliefs. A spirituality different from that of the medieval pilgrim, determined by the Christian interpretation of the world, while nowadays the individual search, the diversity and the experience of the body in motion are the explanatory factors (Creswell and Merriman 2011; Lois et al. 2016).

All of the three motivations that we have described so far are directly associated with the new values of experiential tourism in the twenty-first century. In fact, the Camino is considered a source of health and rest, as it forces an intense daily physical effort in different environments. Walking on foot imitates the exercise done by urban people that go for a run every morning in their place of residence, while the scheduled visit to the gym is replaced by a long walk in the open air. Also culture, always present in trips today, appears repeatedly along a route full of monuments, heritage-listed landmarks, inhabited by rural populations or villages, which retain important identity attributes. In short, in a wide range of recent works we have been able to interpret the Camino de Santiago as a unique, atypical product, but perfectly integrated into the most current motivations of tourism practice.

Finally, doing the Camino also entails many continued satisfactory sensations, which link with the most current leisure practices in developed societies. Among them, enjoying the gastronomy or typical drinks of the places that are crossed, and the celebration of collective meals where pilgrims meet. The gastronomy is highly valued by a walker who, at the same time, suffers the intense physical strain of the route. Without a doubt, the Camino is lived intensely in its integrity, as a space where resistance is put to the test and where simple pleasant sensations are mixed in conversations, over a glass of wine, some cheese or a local vegetable, and the tasting of wines, beers or spirits before bedtime. The experience of the one who walks to Compostela is enriched by learning from the cultures and identities of the villages and towns that are crossed. This evidence constitutes the last element of a particular and satisfactory tourism activity, since attending performances by folk groups and bands or small plays along the Camino, end up building an atypical tourism product but highly seductive, due to its character of total disconnection and openness to different ways of living and expression.

4 The Pilgrim as a New Tourist Who Walks or Rides a Bike

Traditionally, the question of whether the pilgrim is a true tourist or constitutes a special category of person who decides to move or travel has generated much debate (Collins-Kreiner 2010; Frey 1998; Nolan and Nolan 1989; Turnbull 1992). At present, and this is stated in previous works, pilgrims are considered cultural, spiritual or religious tourists with their own attributes (Lois and Lopez 2012). Among them, the aforementioned slow mobility, the importance of having time for reflection or the act of moving along a linear itinerary to a specific destination. But these features do not invalidate the evidence that a pilgrim is a person who spends the night outside his home, who decides to travel during their time off work or other responsibilities, and who seeks a different way to rest and disconnect from everyday life. Therefore, pilgrims are special type of tourists and travellers who can alternate their roles between the historical pilgrimage and the conventional tourist practice. In fact, in several investigations it has been proven that the pilgrims normally become conventional tourists after their arrival in the city of Santiago de Compostela (Lois and Lopez 2012). They enter the city dressed as walkers, they get the *Compostela* and they access the Cathedral, but from that moment on, they book a hotel, hostel or apartment, change their clothes and they have lunch or dinner alone or with others pilgrims with whom they have arrived, visiting bars and shops, and even spending the night out to celebrate that the objective has been achieved.

The transformation of pilgrims into tourists once they get the *Compostela* and go to the Cathedral reaffirms the idea that the Camino de Santiago constitutes a vital parenthesis to recapitulate and, at the same time, allows the verification of the many specific attributes of the contemporary pilgrim. Thus, when deciding to travel the route, many features of a conventional trip are forgone: using a motor vehicle, sleeping in a standard tourist establishment for a number of days, being physically relaxed or permanently seeking enjoyment. In addition, the pilgrim dresses in a special way, as we have already said, influenced by the current sports fashion and trekking equipment, and follows a series of contemporary rites, which vaguely try to evoke the behaviour of Middle Ages walkers. Among them, the preparation of the walking stages and the commitment to record the places that have been visited to obtain the *Compostela*, or the group dinners with other pilgrims, to name a few examples. There is a willingness to assume a role and a series of Jacobean attributes, which always seek to treasure the memory of the uniqueness of the Camino de Santiago experience.

Although all the arguments we have just made are true, the colder and more distant academic analyses affirm that the pilgrim is a type of cultural and/or religious tourist, a person who spends the night outside their habitual residence for several nights with a leisurely motivation. The objectives of the traveller to Santiago are also to know new places and territories, enjoy the landscape, the monumental heritage and eat well, among others. Therefore, the pilgrimage is always classified (somewhat singularised) and is the subject of public policies framed in tourism. Here, the quality of the hotels and the accommodation offer in general, the access or the complementary facilities of the destinations are much less interesting. However, protecting the identity and more ecological and soft character of pilgrim tourism, its characteristics do not diverge from this great economic activity of the current world.

In short, the revitalisation of the Camino defines a particular type of contemporary human being, almost always called a pilgrim, who needs to disconnect from his or her usual daily life and gain time to reflect, even if this does not imply a denial of his or her tourist identity. This particular tourism has created new and attractive places, usually rural areas or small- and medium-sized historical cities, where the route creates a new centrality. Everything crossed or very close to the Camino has been symbolically and economically revalued. New growth territories, linearly articulated and with restored landscapes are created in regions that until recently were characterised by a marked decline in the rural world. The Camino has become a fundamental factor of dynamism in localities that reappear on maps as attractive landmarks that reinforce their value by the deep historicity of the route to Santiago.

5 Conclusion: The Camino as a Linear Axis of Development

Throughout this work we have tried to analyse the contemporary recovery of the Camino de Santiago, its causes and consequences, as a new tourist product filled with originality. The current Compostela route is the result of a *construction* of public administrations, thanks to the return of voluntary pilgrim movements and the need to give coherence to the monumentalisation of a linearly organised European cultural itinerary. Also, the Jacobean pilgrimage strikes for its markedly contemporary character. It is the men and women of the present, subject to daily routines, usually living in cities, believers or not, who star in a massive endeavour on foot or by bicycle to have time for rest, reflection and introspection, and enjoy all the attributes which entail slow travelling. Finally, the pilgrim is seen as a particular type of tourist, who gives up a series of comforts of conventional holiday practice, and who adopts various roles throughout the day when he or she marches towards Santiago and once he or she arrives in Compostela.

From an interpretation centred on the present, it is clear that the Camino de Santiago has managed to generate a linear axis of development in rural areas and small cities that the route crosses. This incontestable evidence is still difficult to specify and quantify, since the benefits of the increase in the number of pilgrims can only be seen in the businesses and the jobs created along the Camino to Santiago, being difficult to disaggregate data locally or by street, since most of them are presented at municipal or parish level. There is an axis of development, still very fine, that is gradually extending towards neighbouring spaces. In addition, and as a final reflection, the physical intervention in the Camino has created a model of architectural action with a certain future, since in the face of the dynamics of urban growth of the past or the limited experiences of rehabilitation, a green and sustainable infrastructure project for hundreds of kilometres offers new opportunities for urban planners, engineers, territorial planners and landscape designers both in the short and medium term.

References

Badone E, Roseman S (eds) (2004) Intersecting journeys. The anthropology of pilgrimage and tourism. University of Illinois Press, Chicago

Bermejo López MB (2001) El Camino de Santiago como bien de interés cultural. Xunta de Galicia, Santiago de Compostela

Castro Fernández BM (2010) O redescubrimento do Camiño de Santiago por Francisco Pons Sorolla. Xunta de Galicia, Santiago de Compostela

Castro Fernández BM (2013) Francisco Pons Sorolla. Arquitectura y Restauración en Compostela (1945–1985). Santiago de Compostela: Consorcio de Santiago and Universidade de Santiago de Compostela

Caucci von Saucken P (1993) El Camino y los Caminos de Santiago. Cuenta y Razón, 76–77 (Special Issue: Camino de Santiago), pp 28–33

Coleman S, Eade J (2004) Reframing pilgrimage. Cultures in motion. Routledge, London and New York

Collins-Kreiner N (2010) The geography of pilgrimage and tourism: transformations and implications for applied geography. Appl Geograph 20(1):153–164. https://doi.org/10.1016/j.apgeog.2009.02.001

Consejo Español de Turismo (2007) Horizonte 2020 del turismo español. Documento base. Consejo Español de Turismo, Madrid

Creswell T (2004) Place, a short introduction. Blackwell Pbs, Oxford

Creswell T, Merriman P (2011) Geographies of motilities. Practices, spaces, subjects. Routledge, London

Falque E (ed) (1994) Historia Compostelana. Clásicos Latinos Medievales. Akal, Madrid

Franco Taboada JA, Tarrio Carroeaguas S (eds) (2000) A arquitectura do Camiño de Santiago, descrición gráfica do Camiño Francés en Galicia. Xunta de Galicia, Santiago de Compostela

Frey NL (1998) Pilgrim stories. On and off the road to Santiago, journeys along an ancient way in modern Spain. University of California Press, Berkeley

García Cuetos MP, Almarcha ME, Hernández-Martínez A (eds) (2010) Restaurando la memoria. España e Italia ante la recuperación monumental de posguerra. Trea, Gijón

Greenia G (2014) What is pilgrimage? In: Harman LD (ed) A sociology of pilgrimage. Embodiment, identity, transformation. Ursus, London, pp 8–28

Lois González RC (2013) The Camino de Santiago and its contemporary renewal: pilgrims, tourists and territorial identities. Cult Relig 14(1):8–22. https://doi.org/10.1080/14755610.2012.756406

Lois González RC, Somoza Medina J (2003) Cultural tourism and urban management in northwestern Spain: the pilgrimage to Santiago de Compostela. Tour Geograph 5(4):446–461. https://doi.org/10.1080/1461668032000129164

Lois González RC, Lopez L (2012) El Camino de Santiago: una aproximación a su carácter polisémico desde la geografía cultural y el turismo. Documents d'Anàlisi Geogràfica 58(3):459–479. https://doi.org/10.5565/rev/dag.6

Lois González RC, Santos Solla XM (2015) Tourist and pilgrims on their way to Santiago. Motives, Caminos and final destinations. J Tour Cultural Change 13(2):149–165. https://doi.org/10.1080/14766825.2014.918985

Lois González RC, Castro Fernández BM, Lopez L (2016) From sacred place to monumental space: mobility along the way to St. James. Mobilities 11(5):770–789. https://doi.org/10.1080/17450101.2015.1080528

Lois González RC, Santos Solla XM, Taboada de Zúñiga MP (2018) The Camino de Santiago de Compostela: The most important historic pilgrimage way in Europa. In: Olsen DH, Trono A (eds) Religious pilgrimage, routes and trails. Sustainable development and management. CAB International, Wallinford (Oxforshire. UK) & Bosto,n pp 72–88

Lois González RC, Lopez L (2019) Liminality wanted. Liminal landscapes and literary spaces: the way of St. James. Tour Geograph 22(2):433–453. https://doi.org/10.1080/14616688.2019.1647452

Lopez L, Lois González RC, Castro Fernández BM (2017) Spiritual tourism on the Way of Saint James. The current situation. Tour Manag Perspect 24:225–234. https://doi.org/10.1016/j.tmp.2017.07.015

Lopez L, Pazos Otón M, Piñeiro Antelo MA (2019) Existe overtourism en Santiago de Compostela? Contribuciones para un debate ya iniciado. Boletín de la Asociación de Geógrafos Españoles (BAGE) 83. http://dx.doi.org/10.21138/bage.2825

López Trigal L (1993) Políticas de rehabilitación de las ciudades y villas camineras de León. In: de Torres MP, Pérez Alberti A, Lois González RC (eds) Los Caminos de Santiago y el territorio. Congreso Internacional de Geografía. Xunta de Galicia, Santiago de Compostela, pp 755–769

Martí Aris C (ed) (1995) Santiago de Compostela: la ciudad histórica como presente. Consorcio de Santiago, Santiago de Compostela

Moralejo Álvarez S (1993) Santiago, Camino de Europa. Culto y cultura en la peregrinación a Compostela. Xunta de Galicia, Santiago de Compostela

Nolan ML, Nolan S (1989) Christian pilgrimage in modern western Europe. University of North Caroline Press, Chapel Hill (NC)

Observatorio do Camiño (2007–2010). Informe do perfil do peregrino. Xunta de Galicia and Universidade de Santiago, Santiago de Compostela

Oficina del Peregrino (1990–2019) Registro de peregrinos llegados a Santiago de Compostela. Series Anuales. Santiago de Compostela: Archicofradía de Santiago

Pichel Pichel JM (2004) Arquitecturas en el Camino. Xunta de Galicia, Santiago de Compostela

Real Decreto 260, 1991 (1991). Real Decreto de 1 de Marzo, sobre organización del Real Patronato de la Ciudad de Santiago de Compostela. (BOE 05-03-1991). Boletín Oficial del Estado, Madrid

Rey Castelao O (2006) Los mitos del Apóstol Santiago. Santiago de Compostela, Nigratrea

Rodríguez Rodríguez M (2004) Los Años Santos compostelanos del siglo XX. Crónica de un renacimiento. Xunta de Galicia, Santiago de Compostela

Somoza Medina J, Lois González RC (2018) Improving the walkability of the Camino. In: Hall M, Ram Y, Shoval N (eds) The Routledge international handbook of walking. Routledge International Handbooks, Oxon, Abingdom (UK), pp 403–413

Soria y Puig A (1993) El Camino a Santiago. 1. Vías, viajes y viajeros de antaño. Ministerio de Obras Públicas y Transporte, Madrid

Torres Luna MP, Pérez Alberti A, Lois González RC (1993) Los Caminos de Santiago y el territorio. Congreso Internacional de Geografía. Xunta de Galicia, Santiago de Compostela

Turnbull C (1992) Postcript: anthropology as pilgrimage, anthropologist as pilgrim. In: Morinis A (ed) Sacred places, sacred spaces. The geography of pilgrimage. Louisiana State University, Baton Rouge, pp 257–274

Turner V (1974) Dramas, fields and metaphors: symbolic action in human society. Cornell University Press, Ithaca

UNWTO (2005). Indicators of Sustainable Development for Tourism Destinations. A Guidebook. UNWTO, Madrid

Urry J, Larsen J (2011) The tourist gaze 3.0. Sage, London

Valle Pérez C (1985) Santiago de Compostela: 1000 años de peregrinación europea. Compostellanum: revista de la Archidiócesis de Santiago de Compostela 30(3–4):479–481

Vázquez de Parga L, Lacarra JM, Uría Ríu J (1948–49) Las peregrinaciones a Santiago de Compostela, vol 3. Consejo Superior de Investigaciones Científicas (CSIC), Madrid

Xunta de Galicia (1993) La Rehabilitación del Camino Francés y la Red de Albergues. Xunta de Galicia, Santiago de Compostela

Xunta de Galicia (1994) Xacobeo. Bibliografía. Xunta de Galicia, Santiago de Compostela

The Success of the Cycle Tourist Backbone Along the Danube in Germany and Austria

Michael Meschik

Abstract The Danube Bicycle Trail consists of several sections which are in different stages of development across Europe, from Germany to Bulgaria/Romania at the Black Sea. The best developed and historically best-documented sections are in Germany and Austria. These can be models for bicycle tourism in other regions. Based on the historical development of the western Danube Bicycle Trail, also known as EuroVelo No. 6, this article presents important basic and advanced requirements for successful bicycle tourism. In the case of the Danube Trail the basic conditions for touristic success already existed, when in the 1980s forward-looking tourism managers and planners at local and regional levels determinedly took steps towards developing the region. Retrospectively, branded touristic routes like the Danube Bicycle Trail assisted in conserving regional ecosystems and heritage, while at the same time opening economic opportunities for local businesses and generating jobs. For continuous success constant improvements to safeguard a leading position as an outstanding tourism region are necessary. Other regions seeking to emulate the accomplishments of the Danube Trail should also highlight their specific heritage and natural assets.

1 The Danube Bicycle Trail

Touristic bicycle routes along rivers create win–win situations. Bicycle tourists enjoy the gentle ride following the downhill gradient. In Alpine regions it is common practice to cycle downstream, only a few tourists pedal upstream. Local politicians also favour routes along rivers, as there are usually service roads along river banks, motorized traffic is mostly restricted and the utilization for bicycle traffic can normally be achieved more easily and cheaper than elsewhere. As rivers have historically been major thoroughfares, we also find numerous ancient buildings and many other sights

M. Meschik (✉)
University of Natural Resources and Life Sciences, Vienna, Austria
e-mail: michael.meschik@boku.ac.at

P. Pileri and R. Moscarelli (eds.), *Cycling & Walking for Regional Development*, Research for Development, https://doi.org/10.1007/978-3-030-44003-9_16

near them. All this is ideal for bicycle tourism, practically the most prominent category in "ecotourism". There are several definitions of ecotourism, the definition of the (UNWTO 2019) seems well suited:

Ecotourism is a type of nature-based tourism activity in which the visitor's essential motivation is to observe, learn, discover, experience and appreciate biological and cultural diversity with a responsible attitude to protect the integrity of the ecosystem and enhance the well-being of the local community. Ecotourism increases awareness towards the conservation of biodiversity, natural environment and cultural assets both among locals and the visitors and requires special management processes to minimize the negative impact on the ecosystem (UNWTO 2019).

The bicycle route along the Danube is regarded as one of international renown. The Danube Bike Trail has a total length of 2850 kilometres. From the springs in Donaueschingen, Germany, it connects Austria, Slovakia, Hungary, Croatia, Serbia, Bulgaria and Romania and it ends near the Danube estuary (Tulcea) at the Black Sea. The trail follows ancient Roman roads (Via Istrum), basically the former northern border of the Roman Empire in this region (Wikipedia 2019b). As the Danube has been a border region for centuries, we also find innumerable fortifications, monasteries and historical sites along its course.

EuroVelo, the European cycle route network "... *was initiated by the European Cyclists' Federation ECF to develop a network of high-quality cycling routes linking all countries in Europe. It can be used by long-distance cycle tourists as well as by local people making daily journeys*" (ECF 2019). Effective October 2019, the EuroVelo project included 19 trans-European cycle routes "... *totalling over* 70,000 km, *of which about* 45,000 km *is already in place*". Except for the first stage in Germany from its origin to Tuttlingen, the Danube Bicycle Trail is also listed as EuroVelo 6, the route from the Atlantic to the Black Sea (Fig. 1). The difference in the length figures given in the ECF documentation also shows a significant problem: Many of Europe's main bicycle trails are projects, still in the planning stage. The ECF distinguishes several stages of EuroVelo route quality, from "(1) *Certified EuroVelo Route*", "(2) *Developed route with EuroVelo signs*", "(3) *Developed route*"

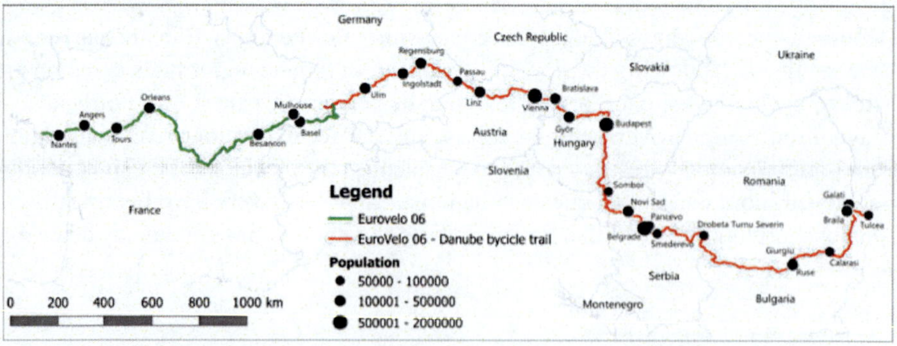

Fig. 1 The EuroVelo 6 crosses Europe from the Atlantic to the Black Sea. Except for the first few kilometres, the Danube Trail makes up the middle and eastern two-thirds of EuroVelo 6

until "(4) *Route under development*" and finally "(5) *Route at the planning stage*". The Danube Route is well established from its origin in Germany through Germany, Austria and Slovakia, Hungary and some parts in Serbia. In Romania, Bulgaria and Serbia—specifically outside cities—the route shares motorized traffic on sometimes busy and dangerous roads, lacks signage to a great extent and offers hardly any tourism infrastructure like campsites, let alone hotels or guest houses (Bondarev 2019). From Tuttlingen, Ulm, Ingolstadt, Regensburg to Passau in Germany, the EuroVelo 6 route is graded (2) by the ECF, in Austria (Linz, Krems, Vienna) (2, 3), in Slovakia (Bratislava) (2, 3), in Hungary (Györ, Komaron, Budapest) (2, 3), in Serbia (Belgrade) (2), in Bulgaria and finally in Romania (4). Bicycle tourists, however, find the south-eastern parts of EuroVelo 6 quite difficult, complaining of a (total) lack of signage and many stretches of the route along busy roads in mixed traffic with motor vehicles (Bondarev 2019). It is therefore necessary to look at the different sections of the Danube Bicycle Trail individually. The success story of the Danube Bicycle Trail as an outstanding ecotourism route is definitely true of the sections in Germany and Austria and dealt with in the following chapters.

2 The Danube Trail in Germany and Austria

Today, a vast multitude of touristic websites informs about all the amenities and must-see sights along the German and Austrian sections of the Danube Bicycle Route (see Fig. 2). In some cases, the web information is quite superficial, but in most cases the details given are strikingly detailed and up-to-date. Along the historical route there are many small villages as well as some European capitals. On the one hand, towns and cities are necessary for accommodation and cultural highlights, on the other hand cultivated rural landscapes and well-preserved nature are key aspects of

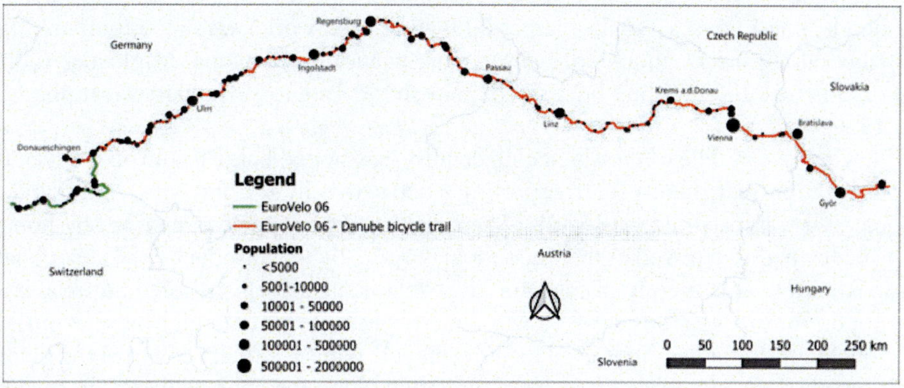

Fig. 2 The Danube Bicycle Trail in Germany and Austria (EuroVelo 6)

an attractive route. The Danube boasts several natural reserve areas along its course (Natura 2000, national parks, etc.).

According to the German Cyclists' Federation ADFC, the German part of the Danube Trail ranks number four among all German routes (ADFC 2019) and the international stretch of the Danube Trail is by far the most popular route outside Germany (Travelbike Vermietung and ADFC 2019, p. 32). How did the Danube Bicycle Trail become so well known and successful?

2.1 Historical Development

Scoping old references soon leads to the insight that "*How it all began ...*" is not part of the story of bicycle tourism along the Danube. Nobody "invented" the Danube as a well-known touristic route. Its development seems to have been a long process, initiated by some forward-looking individuals: Having established a leading position in bicycle tourism, persons responsible (in tourism) are motivated to keep these standards in a highly competitive market by constant efforts. To understand current well-established and recognized standards, it is interesting to analyse the historical development of the Danube Bicycle Route.

Big rivers have always been shipping lanes. Towards the end of the eighteenth century, regular passenger-ship traffic was established on the Danube and became affordable also for ordinary people. In 1829 the first steamboat travelled from Vienna to Budapest. In 1854 the German princess "Sisi" took the steamboat "Regensburg" from Passau to Linz, where Emperor Franz Josef was awaiting his bride. Since the Biedermeier period (roughly 1815–1848), the upper middle class from Vienna and other cities of the Austrian-Hungarian monarchy (Budapest, Bratislava) enjoyed their summer retreat ("Sommerfrische") in the Austrian Upper Danube valley, taking in the striking countryside and historical background. The more affluent families had mansions and brought their entourage with them, the less well-off visitors stayed at hotels, guest houses, etc. In the period between the world wars (eco)tourism flourished along the Danube—focusing on the beautiful landscape. Miglbauer (2009) reports that 2000 folding boat riders paid short visits to Engelhartszell in the year 1929.

Until the middle of the nineteenth century, eastbound barge traffic on the Danube used the current, mainly for transporting wood to the cities. Corn, wine, ironmongery, etc., were transported westwards. The barges had to be towed upstream by horses. The towpaths, deprived of their historical purpose by the introduction of steamboats, were afterwards merely needed for river maintenance. This offered an invaluable opportunity to instal attractive bicycle paths on them for the emerging bicycle tourism. Inhabitants of larger cities used the new bicycle paths for (half) day trips (originally illegally), initially proceeding from the cities. Soon the gaps between the sections around the cities were closed; towpaths and also river dam crests around hydropower stations built since the 1970s, abandoned rail tracks and low volume roads were connected. On some stretches also (busy) streets shared with motorists had to be included in the trails.

Before the private car started its still ongoing triumph in the 1970s, over a million people used the bicycle in Austria for their daily mobility; over 200,000 in Vienna, a third of them women. In the 1950s, the bicycle was the prevalent vehicle on the streets. In the 1960s and 1970s, using one's private car for the journey to the holiday destination became the new feeling of personal freedom. At the same time, using the bicycle as sole means of transport on vacation became a new statement of the "back-to-nature movement". The growing numbers of cars demanded motorways bypassing the roads traversing historical places and regions. When the first Austrian motorways, like the Innkreis motorway between Passau and Linz opened (1979), the motorized through-traffic in towns and villages along the Danube decreased considerably, consequently the accommodation providers and restaurant proprietors lost their customers. This loss was soon compensated by the cyclists. The leading motive for bicycle holidays is nowadays the quiet, relaxing mobility off the motorized traffic routes. Travelling by car is easy and fast, but also boring and impersonal. Crossing the Alps on a bicycle or carrying all your luggage in panniers are some achievements bicyclists seek.

Historical details on the Danube Trail in Germany and Austria (in German) are well documented in Wikipedia (2019a). More detailed information gives Miglbauer (2009). Miglbauer rates the Danube Trail as a trendsetter for international bicycle tourism. As an expert in practical bicycle touristic development he locates the origins of European bicycle tourism above all in Austria and Germany. Some people in the Austrian provincial state of Upper Austria started to see bicycle tourism along the Danube—along with adjacent bike paths in the region of "Salzkammergut" and along the rivers Inn and Enns—as a welcome opportunity to stimulate economic development in a less prosperous region. Almost at the same time the provincial states of Bavaria in Germany and of Lower Austria in Austria started to promote bicycle tourism. Delving for the historical origins, Miglbauer scrutinized all sources he could obtain, like studies, documents, newspaper articles, annual reports of tourist associations, etc. He even conducted several interviews, discussions and focus groups with tourism people back in 2008. In his book he lists 32 persons from different organizations (see Table 1). This list is interesting, as the different stakeholder organizations involved in bicycle tourism can be clearly discerned.

The origins of bicycle tourisms (where, when, who and how) cannot be traced exactly but are ascribed to the "*first half of the* 1980s". Touristic innovation cannot be planned entirely, it is also demand-driven by the market. Miglbauer quotes Georg Steiner, head of the tourism association of the city of Passau from 1982 to 1989, who said "*bicycle tourism arrived soft-footed*" (Miglbauer 2009, p. 14). Table 2 lists events important for cycle tourism along the Danube.

Danube tourism in the 1970s was initially boat tourism (folding canoes). Optimistic tourism managers had hoped that many boaters would stay overnight and had invested in new accommodation accordingly. However, the boaters seldom stayed overnight. It was sheer luck that the cyclists came for food and drink and they also

Table 1 Persons interviewed in search of the touristic origins of bicycle tourism along the Austrian and Bavarian Danube (condensed from Miglbauer 2009)

Responsibility, organization or enterprise	
Tourism association, tourism managers	12
Inns, hotels, guest houses	11
City tour officials, culture management	2
Planning total traffic, public transport (Upper Austria)	2
Austrian federal railways	2
German Cyclists' Federation ADFC	1
Bicycle tour operator	1
Danube shipping company	1
Total number of people interviewed	**32**

Table 2 Touristic developments of the Austrian/Bavarian Danube (and Inn) regions in the rise of bicycle tourism (A short chronology, condensed from Miglbauer 2009 and other sources)

Date	Event	Sources
1954	The first tourism association was founded in Engelhartszell	Miglbauer (2009)
1979	Foundation of "Werbegemeinschaft oberes Donautal"	
1980	Along the Altmühl, a feeder river of the Danube, about 30 kilometres of abandoned railroad tracks between Eichstätt and Beilngries were adapted as a bicycle trail and signposted	
1980s	1979 the meadows along the river Inn from Hochburg/Ach to Passau (55 km) were installed as a European Protected Reserve. In the middle of the 1980s a joint Bavarian/Austrian section of the trail was equipped with road signs	
1982	Opening of the first kilometres signposted Danube Bicycle Trail (Landshaag, left bank)	
1982	The first kilometres of bicycle trails on towpaths along the Danube near Ottensheim are "illegally" signposted. The first organized bike ride attracts 1,000 participants	Miglbauer, in a talk with Traunmüller, Nov. 25, 2019
1983	Cycling is officially approved on towpaths near electric power stations; consistent signposting follows	

(continued)

Table 2 (continued)

Date	Event	Sources
1984	Hans Hofer, member of the Austrian parliament, suggests the development of the towpaths along the Danube between Schlögen and Aschach for bicycle tourism. Mr. Sekanina, then minister of economic affairs, denies funding due to scarcity of budget	Neue Kronenzeitung, Jan. 27, 1985, cited in Miglbauer (2009)
1984	Manfred Traunmüller publishes the first bicycle touring guide for the Danube Bicycle Trail in Upper Austria and the adjoining areas	Traunmüller (1984)
1985	As a consequence of the popular petition against the Danube hydropower station in Hainburg (downstream Vienna) in spring 1985, the plans for Hainburg were relinquished (the final stop came on July 1, 1986). Consequently, money for the Upper Austrian towpaths was disposable. The money (ATS 14 million, the equivalent of one million Euro) was used for a "*generous redevelopment of the towpaths*" towards a "*unique El Dorado for cyclists*". Intended was "*a substantial impetus for the touristic regions Upper Danube valley and the region around Eferding*"	Neue Kronenzeitung, Jan. 27, 1985, cited in Miglbauer (2009)
1985	Paul Pollak publishes the first bicycle touring guide for the Danube Bicycle Trail in Austria from Passau to Hainburg	Pollak (1985)
1986	The 300th birthday anniversary of the German Baroque artist Cosmas Damian Asam is also commemorated with a bicycle route, the "Tour de Baroque" from Neumarkt to Passau	Miglbauer (2009)
1988	The Austrian provincial state of Upper Austria systematically develops and publishes plans for state-wide bicycle routes	
1989	The first bicycle touring guide for the Inn Bicycle Trail (along the border between Austria and Germany) is published. Cyclists were asked "*not to leave the marked trail, and thus not to disturb animals or destroy plants*"	
1990	The first tourist survey on the Danube Bicycle Trail is carried out (Schuller, Miglbauer). Surprisingly, bicycle tourists are well educated and well off	Miglbauer and Schuller (1991), Dreyer et al. (2012) p. 184

(continued)

Table 2 (continued)

Date	Event	Sources
1996	The tourism organization "Werbegemeinschaft Donau" establishes "top-bicycle-stops", an initiative to ensure certified high-quality accommodation for tourists	Miglbauer (2009)
2008	The tourism platform "ARGE Donau Österreich" unites several former "Werbegemeinschaften" along the Austrian Danube	Dreyer et al. (2012)
2010	The "Werbegemeinschaft Donau" lists 26 top-bicycle-stops along the Austrian Danube	

appreciated the new rooms. This motivated several accommodation providers to invest hope and money in bicycle tourism.

Guests on cycling tours were different from other tourists previously staying: They now wanted to stay for only one night, whereas other guests had usually stayed for two or even three weeks in their summer retreat. The overnight-only stays meant a drastic change of routines for accommodation providers. The bicycle tourists still stay several nights in the region but in different accommodations. Most cycle tourists spend one week on the Danube Trail from Passau to Vienna (Meschik 2012).

The first bicycle tourists were quiet and modest, inquisitive and interested in first-hand information on the next day's route, accommodation, ferry links, must-see sights, etc. They were characterized as "*upper middle-class*", seeking "*relaxed days as alternatives from their white-collar office jobs". Their "unpretentiousness regarding apparel and looks irritated many a host in the beginning, the more as they asked for high quality accommodations*" (Miglbauer 2009).

2.2 Current Situation

There is an ongoing trend for shorter holidays nowadays, from several weeks in the 1970s to one week or less. The average time spent by tourists in Austria was 3.4 days in 2016 (Statistik Austria 2019). Tourism regions and travel agencies offer numerous options and bicycle tourism fills a niche in the tourism market. Competition between growing numbers of bicycle tourism destinations is also increasing. Several types of bicycle-related vacations have emerged: Day trips, classical bicycle journeys of about a week, mountain biking, racing bike tours, pilgrimages, etc. The use of power-assisted bicycles (pedelecs – pedal electric cycles) is increasing steadily. In several European countries, among them Austria, one out of three bicycles sold is a pedelec (Leschner 2019).

Miglbauer stresses that it is important for (bicycle) tourism to constantly review the steps already taken and to preview and explore new horizons. Bicycle tourism is no fast-selling business. The competitive market needs permanent efforts, improvements and an innovative spirit, also to anticipate new trends before competitors do so Dreyer et al. (2012).

The national economic benefit of cycling for Austria was estimated for 2008 as € 600 million plus 10,800 jobs (full-time equivalent). Bicycle tourism generates 53% of this revenue and 70% of the jobs. The Austrian Danube Trail earns 12% of the tourism benefit which accounts for € 38 million and 900 jobs (Miglbauer et al. 2009).

The permanent improvement of the infrastructure of the Danube Trail (118 km) and the Inn trail (55 km) cost approximately € 20 million in Upper Austria alone until 2007. In 2006 a joint Bavarian/Austrian suspension bridge for pedestrians and cyclists with a span of 144 m, the "Mariensteg", was built for € 1.6 million over the Inn between Wernstein and Neuburg (Miglbauer 2009). As of May 2014, the provincial state of Lower Austria reported a sum of € 28 million, which was invested in the Danube Bicycle Trail in the "*last thirty years*". These investments result in a "*yearly gross revenue of ten million Euro*"; 80% of the sum is attributed to overnight guests (Bohuslav 2014). Figure 3 shows how multifaceted the Danube Trail is.

Since the first tourist surveys on the Danube Bicycle Trail, the cyclists interviewed have had a higher education than the average traveller. Bicycle tourists still expect scenic attractions, increased fitness and active mobility, non-motorized holidays (away from the car), variation and diversion, cultural and historical highlights, independence, etc. Some of these reasons for cycling show "*post-materialistic features*" (Dreyer et al. 2012; Zeitlhofer 2001).

Since 2000, automated permanent counters have been installed along the Danube, in Austria at 17 locations. From these we know that about 670,000 cyclists ride along the Austrian Danube every year. Surveys with bicycle tourists are conducted every five to six years (ARGE Donau Österreich 2018). Some authors assume that bicycle tourism along the Danube has reached its zenith already. As the total number of cyclists is stagnating, there seem to be fewer tourists and more day-trippers and accommodation providers report a shorter season and increasing dependence on the weather nowadays.

2.3 Quality Standards and Certifications

Each country along the Danube—sometimes even provinces—have their guidelines on how to design bicycling infrastructure. In Austria these are the Planning Guidelines for Bicycle Traffic (FSV 2014), in Germany the Recommendations for Bicycle Infrastructure (FGSV 2010). A well-renowned certification standard for touristic bicycle routes was issued by the German bicyclists' association (ADFC 2014a), for

Fig. 3 Different sights along the Danube Trail, from top left, clockwise: historical track in the Wachau | Krems—Stein | Hainburg, downstream Vienna | Linz | near Korneuburg | Ybbs

entire tourism regions (ADFC 2014b). Moreover, the ECF has issued a certification standard for the EuroVelo routes (Bodor et al. 2018). Local tourism organizations also try to establish local certification standards, for example the "bed&bike cachet" (Bett&Bike Gütesiegel). In 2019, 32 bed&bike accommodation providers were certified in Lower Austria (Donau Niederösterreich Tourismus GmbH 2019). Certified bed&bike providers offer one-night accomodation, provide safe bicycle storage, clothes drying facilities, tools for basic repairs, and offer warm meals all day.

In an ongoing European Danube transnational research programme (Interreg) special guidelines are currently developed. These guidelines aim at standardizing minimum "must have" and commendable "nice to have" criteria for ecotourism and bicycle tourism in particular (Aschauer et al. 2019). The goal is to support aspiring bicycle tourism destinations without discouraging them with too high standards already being used in well-established destinations as along the Danube in Germany and Austria. Among other important requirements already mentioned above, these guidelines also define the intermodal connection of the bicycle trip with public transport as a feeder mode to connect to the starting point and destination of the bicycle tour. The chapters of these guidelines are shown in Table 3. A checklist for implementation is currently in the testing phase. Success stories and best practice examples should enable all potential regions and organizations to develop bicycle tourism successfully. The authors regard the Danube Bicycle Trail as an apt model for developing bicycle tourism.

Table 3 Chapters of the guidelines for sustainable bicycle tourism (Aschauer et al. 2019)

Chapter	Content
Bicycle-based ecotourism	Highlighting the benefits for local economy, nature conservation and fostering of local culture and heritage
Planning for different types of cyclists	Basic design principles; vehicular speed and traffic volumes; recommending segregated infrastructure
How to start bicycle tourism in your region	Advertising and marketing natural beauty and regional characteristics; thematically branded touristic routes
Infrastructure for high-level bicycle tourism	Main infrastructure elements worth knowing in bicycle traffic; defining "must have" and "nice to have" criteria for improvements; seeking advice of skilled planners
Transport services and intermodality	How to reach the starting point of the tour and how to return home with public transport; ecotourism requires the use of environmentally friendly modes
Bicycle rental schemes	Leave your own bicycle at home; get one from your accommodation host or use a public shared bicycle
Accommodation and gastronomy	These two have to work hand in hand; overnight stays and gastronomy return the most profit
Information, communication and marketing	Information necessary to attract bicycle tourists; success factors to properly place "cycling tourism" on the market
Measuring bicycle-based ecotourism	Frequent counts of cyclists and opinion surveys reveal characteristics; cyclist's feedback helps in permanent improvement

Fig. 4 EuroVelo 6 "Donauradweg" approaching Vienna; adjoining routes, service facilities and sights are illustrated in this overview map (© arbeitsgemeinschaft kartografie)

2.4 *The Danube Trail as a Role Model*

Bicycle tourism is an important mobility form supporting ecotourism. To be successful, bicycle routes must fulfil certain quality standards. The bicyclists increasingly plan their tours online; consequently, they want to see what to expect on route. Bicycle tourism along the Danube Trail also depends on marketing in all possible media.

The Danube Bicycle Trail was initiated by some forward-looking individuals and is now well embedded in governmental transport and tourism plans, including adequate funding. Meanwhile it has a tradition of over 30 years. Along the trail, tourists can discover several adjacent routes and sights (Figs. 4 and 5). The promotion of bicycle tourism has shown benefits for local economy and nature conservation, also fostering local culture and heritage. The historical developments and the quality standards achieved are suitable standards for developing new regions and trails for bicycle tourism in most comparable circumstances.

Fig. 5 Burg Kreuzenstein—one of many sights on an adjoining route of EuroVelo 6 (Picture by courtesy of F. Kluibenschädl)

Acknowledgements This article honours the profound knowledge of Ernst Miglbauer and Manfred Traunmüller. The Danube Trail was initiated by a handful of local stakeholders rather than by governmental transport authorities. Miglbauer identifies Traunmüller (former tourism manager, now one of the leading bicycle tour operators in Europe) as its originator, along with Walter Steiner (former mayor of Ottensheim) and Paul Pollak (journalist), all of them avid cyclists and still active.

References

ADFC (2014a) Klassifizierung von ADFC-Qualitätsradrouten. In (Vol. 4. Auflage). Allgemeiner Deutscher Fahrrad Club (ADFC), Berlin

ADFC (2014b) Zertifizierung von ADFC-RadReiseRegionen. In (Vol. 4. Auflage). Allgemeiner Deutscher Fahrrad Club (ADFC), Berlin

ADFC (2019) ADFC-Travelbike Radreiseanalyse 2019; Kurzbericht. In. ADFC, Berlin

ARGE Donau Österreich (2018) Radverkehrszählung am Donauradweg 2017 [Press release]

Aschauer F, Gauster J, Hartwig L, Klementschitz R, Meschik M, Pfaffenbichler P, Unbehaun W (2019) WP3 Ecotourism planning—Guidelines for sustainable bicycle tourism—Deliverable 3.3.1; Interreg Danube Transnational Programm, EcoVeloTour, Vienna. http://www.interreg-danube.eu/uploads/media/approved_project_output/0001/36/cfb8514ba973699218d4caf1fb4ec0371766d5d9.pdf. Accessed 10 Dec 2019

Bodor A, Buczyński A, Fahrenkrug E, Freire J, Lancaster E, ADFC Allgemeiner Deutscher Fahrrad-Club e. V (2018) EuroVelo—the European cycle route network—European Certification Standard; Handbook for route inspectors. https://pro.eurovelo.com/download/document/ECS-Manual-2018_04_16.pdf; https://pro.eurovelo.com/download/document/European-Certification-Standard-Manual-short-version-English.pdf. Accessed 10 Dec 2019

Bohuslav P (Producer) (2014, Nov 04, 2019). Donauradweg feiert 30-Jahr-Jubiläum. http://www.ots.at/presseaussendung/OTS_20140507_OTS0145. Accessed 10 Dec 2019

Bondarev D (2019) Discover EuroVelo. Paper presented at the EcoVeloTour—Second Teaching Interaction, Tulcea, Romania. https://steemit.com/travel/@discovereurovelo/discover-eurovelo-6-the-rivers-route. Accessed 10 Dec 2019

Donau Niederösterreich Tourismus GmbH (Producer) (2019, Nov 04, 2019) Am Donauradweg in der niederösterreichischen Donauregion; Radgenuss an der Donau. https://www.donau.com/fileadmin/user_upload/Bilder_und_PDFs/Presse/2019/PT_Donauradweg_2019.pdf. Accessed 10 Dec 2019

Dreyer A, Mühlnickel R, Miglbauer E (2012) Radtourismus: Entwicklungen. Walter de Gruyter, Potentiale, Perspektiven

ECF (2019) EuroVelo. https://ecf.com/projects/eurovelo; http://www.eurovelo.org/; https://en.eurovelo.com/. Accessed 10 Dec 2019

FGSV (2010) Empfehlungen für Radverkehrsanlagen (ERA). In: Köln: Forschungsgesellschaft für Straßen und Verkehrswesen (Hrsg.)

FSV (2014) RVS 03.02.13 Radverkehr. In: Richtlinien und Vorschriften für das Straßenwesen (RVS). Wien: Österreichische Forschungsgesellschaft Straße—Schiene—Verkehr (FSV)

Leschner S (Producer) (2019, Nov 04, 2019) Fahrradverkauf in Österreich so hoch wie seit 10 Jahren nicht mehr. APA OTS press releases. https://www.ots.at/presseaussendung/OTS_20190411_OTS0018/fahrradverkauf-in-oesterreich-so-hoch-wie-seit-10-jahren-nicht-mehr. Accessed 10 Dec 2019

Meschik M (2012) Sustainable cycle tourism along the Danube Cycle Route in Austria. Tour Plan Dev 9(1):41–56. https://doi.org/10.1080/21568316.2012.653478

Miglbauer E (2009) Radgenuss am Fluss; Eine Europäische Erfolgsgeschichte an Donau und Inn. OÖ Tourismus, Werbegemeinschaft Donau Oberösterreich, TV Ostbayern, Linz

Miglbauer E, Pfaffenbichler PC, Feilmayr W (2009) Kurzstudie Wirtschaftsfaktor Radfahren. Die volkswirtschaftlichen Auswirkungen des Radverkehrs in Österreich. https://www.bmnt.gv.at/umwelt/luft-laerm-verkehr/verkehr/radfahren/WirtschaftsfaktorRad.html. Accessed 10 Dec 2019

Miglbauer E, Schuller E (1991) Wie reisen Radler?—Ergebnis einer wissenschaftlichen Untersuchung des Donau-Radweg-Tourismus. Paper presented at the Fachtagung "Fahrradtourismus—eine neue Reiseform?", Rosenheim

Pollak P (1985) Der Donauradweg. Von Passau bis Wien in 10 Etappen—mit einem Anhang von Wien bis Hainburg

Statistik Austria (2019) Regionale Entwicklungen des Tourismus. http://www.statistik-austria.at/web_de/services/wirtschaftsatlas_oesterreich/oesterreich_und_seine_bundeslaender/025911.html. Accessed 10 Dec 2019

Traunmüller M (1984) Meine Radfahrt an der Donau: ein Führer über die Radwege an der oberösterreichischen Donau und dem angrenzenden Gebiet;[ein Reisebegleiter mit Tourenkarte 1: 50.000]: Oberösterr. Landesverlag

Travelbike Vermietung and ADFC (2019) ADFC-Travelbike-Radreiseanalyse 2019; 20. bundesweite Erhebung zum fahrradtouristischen Markt. Berlin. https://www.adfc.de/artikel/adfc-radreiseanalyse-2019/

UNWTO (2019) UNWTO Tourism Definitions| Définitions du tourisme de l'OMT| Definiciones de turismo de la OMT, Madrid. https://www.e-unwto.org/doi/abs/10.18111/9789284420858. Accessed 10 Dec 2019

Wikipedia (2019a) Donauradweg (D6). https://de.wikipedia.org/wiki/Donauradweg_(D6). Accessed 10 Dec 2019

Wikipedia (2019b) The Danube. https://en.wikipedia.org/wiki/Danube. Accessed 10 Dec 2019

Zeitlhofer H (2001) Konzeption eines Erhebungsverfahrens für den touristischen Radverkehr in der Wachau. (Master). Universität für Bodenkultur, Wien

Useful links—Danube Bicycle Trail (all Dec. 10, 2019): Austrian section, interactive map

https://www.donau-oesterreich.at/en/danube-cycle-path/
https://www.donauregion.at/en.html
https://www.austria.info/uk/things-to-do/cycling-and-biking/danube-cycle-path
Germany to Budapest: https://www.danube-cycle-path.com/
Wikipedia: https://de.wikipedia.org/wiki/Donauradweg_(D6)